Istvan M. Mandity

Design of β-peptide foldamers by stereochemistry and topology

Istvan M. Mandity

Design of β-peptide foldamers by stereochemistry and topology

The effect of backbone and side-chain on 3D structure of artificial self-organizing systems

LAP LAMBERT Academic Publishing

Impressum / Imprint

Bibliografische Information der Deutschen Nationalbibliothek: Die Deutsche Nationalbibliothek verzeichnet diese Publikation in der Deutschen Nationalbibliografie; detaillierte bibliografische Daten sind im Internet über http://dnb.d-nb.de abrufbar.
Alle in diesem Buch genannten Marken und Produktnamen unterliegen warenzeichen-, marken- oder patentrechtlichem Schutz bzw. sind Warenzeichen oder eingetragene Warenzeichen der jeweiligen Inhaber. Die Wiedergabe von Marken, Produktnamen, Gebrauchsnamen, Handelsnamen, Warenbezeichnungen u.s.w. in diesem Werk berechtigt auch ohne besondere Kennzeichnung nicht zu der Annahme, dass solche Namen im Sinne der Warenzeichen- und Markenschutzgesetzgebung als frei zu betrachten wären und daher von jedermann benutzt werden dürften.

Bibliographic information published by the Deutsche Nationalbibliothek: The Deutsche Nationalbibliothek lists this publication in the Deutsche Nationalbibliografie; detailed bibliographic data are available in the Internet at http://dnb.d-nb.de.
Any brand names and product names mentioned in this book are subject to trademark, brand or patent protection and are trademarks or registered trademarks of their respective holders. The use of brand names, product names, common names, trade names, product descriptions etc. even without a particular marking in this works is in no way to be construed to mean that such names may be regarded as unrestricted in respect of trademark and brand protection legislation and could thus be used by anyone.

Coverbild / Cover image: www.ingimage.com

Verlag / Publisher:
LAP LAMBERT Academic Publishing
ist ein Imprint der / is a trademark of
OmniScriptum GmbH & Co. KG
Heinrich-Böcking-Str. 6-8, 66121 Saarbrücken, Deutschland / Germany
Email: info@lap-publishing.com

Herstellung: siehe letzte Seite /
Printed at: see last page
ISBN: 978-3-659-58996-6

Zugl. / Approved by: Szeged, University of Szeged, Diss., 2010

Table of Contents

Abbreviations

3D	three-dimensional
ACHC	2-aminocyclohexanecarboxylic acid
ACPC	2-aminocyclopentanecarboxylic acid
APC	4-aminopyrrolidine-3-carboxylic acid
B3LYP	Becke, three-parameter, Lee-Yang-Parr exchange-correlation functional
Boc	*tert*-butoxycarbonyl
Bzl	benzyl
CIP	Cahn-Ingold-Prelog
D	aspartic acid
DBU	1,8-diazabicyclo[5.4.0]undec-7-ene
DCC	dicyclohexylcarbodiimide
DCM	dichloromethane
ΔE	conformational energy difference
DIPEA	N,N-diisopropylethylamine
DLS	dynamic light scattering
DMAP	4-(N,N-dimethylamino)pyridine
DMF	N,N-dimethylformamide
DTT	DL-dithiothreitol
ECD	electronic circular dichroism
EDCI	1-ethyl-3-(3-dimethylaminopropyl)carbodiimide
EDT	1,2-ethanedithiol
Fmoc	9-fluorenylmethoxycarbonyl
H	histidine
HATU	O-(7-azabenzotriazol-1-yl)-N,N,N′,N′-tetramethyluronium hexafluorophosphate
HBTU	O-(benzotriazol-1-yl)-N,N,N′,N′-tetramethyluronium hexafluorophosphate
HOBT	hydroxybenzotriazole
hS	homo-serine
ISPA	Isolated spin pair approximation
K	lysine
MBHA	4-methylbenzhydrylamine
MC	Monte Carlo
MD	molecular dynamics
MOE	molecular operating environment
MS	mass spectrometry
NMP	1-methyl-2-pyrrolidinone
NMR	nuclear magnetic resonance
NOESY	nuclear overhauser effect spectroscopy
PCM	polarizable continuum model

PEC	potential energy curve
PPI	protein-protein interactions
RHF	spin-restricted Hartree-Fock theory
ROESY	rotating frame Overhauser effect spectroscopy
RP-HPLC	reversed-phase high-performance liquid chromatography
RSA	restrained simulated annealing
SPA	stereochemical patterning approach
SPPS	solid-phase peptide synthesis
TAR	transcriptional activator-responsive element
^{t}Bu	*tert*-butyl
TEM	transmission electron microscopy
TFA	trifluoroacetic acid
	fluoro-N,N,N′,N′-tetramethylformamidinium
TFFH	hexafluorophosphate
TIS	triisopropylsilane
TOCSY	total correlation spectroscopy
VCD	vibrational circular dichroism

1. Introduction and aims

Biopolymers governing the functions of life include peptides, proteins, RNA and DNA. They consist of α-amino acids and nucleotides. Nature has developed a high diversity of proteins for different functions, such as enzymes, receptors, transport proteins, etc. To carry out their roles, they need to fold into well-defined hierarchical 3D structures. Foldamers, non-natural self-organizing biomimicking systems, exhibit similar properties to those of proteins, e.g. they have a tendency to fold into specific periodic compact structures.[1-9]

Foldamers have the potential to achieve structural versatility similar to that of the natural proteins, and consequently they have numerous promising biological applications where the tailored 3D structure of the designed foldamers is crucial. Their most important applications are as: antibacterial or cell-penetrating amphiphiles, inhibitors of fat and cholesterol absorption, RNA-binding oligomers and antagonists of cancer-related proteins.[9] Foldamers based on appropriate structural features and pharmaceutical applications can be designated protein mimics.

In pharmaceutical applications, foldamers play roles as a novel class of drug scaffolds with tailored molecular shape and surface. This raises the need for more versatile frameworks with the aim of designing foldamers that bind to almost any surface. The most thoroughly studied representatives of this field are the β-peptides, consisting of β-amino acids.[1-9] They populate a wide range of tunable secondary structures (numerous helices, sheet-forming strands and turn motifs) with a propensity to associate into tertiary structure-like motifs.[4] Novel secondary structural motifs can be gained (*i*) by the use of new non-natural amino acids and derivatives or (*ii*) through an understanding of the structure-promoting effect of the stereochemistry.

Our major aim was to harness the latter understanding to gain new structures with a designed specific backbone configuration. The basic secondary structures for β-peptide foldamers have been established mainly for uniformly substituted β-amino acids with respect both to the side-chain position (β_3 and/or β_2) and to the backbone stereochemistry, leading to a uniformly substituted backbone pattern. These structures composed of alicyclic β-amino acids with *trans* configuration form a helical conformation, while if the geometry is switched to *cis*, the prevailing secondary structure will be strand-like.[4] Seebach *et al.* introduced the alternating design principle to the field and synthesized β_2/β_3 sequences by the sequential coupling of β_2- and β_3-amino acids.[6] These oligomers form the alternating H10/12 helices that are unique to the β-peptides and postulated on the basis of computational calculations to be the most stable secondary structure. It is noteworthy that these oligomers also have important pharmaceutical applications. The alternating coupling of different amino acids has been successfully extended to combined α/β-peptides, which show signs of a propensity to fold for short sequences, even with β-amino acids with alicyclic or sugar-based side-chains.[8] The described secondary structures that exist for β-peptides can be strongly controlled via the side-chain chemistry, and especially the backbone configuration provides a powerful toolkit with which to tailor helical structures. We set out to combine the alternating design principle with the stereochemical configuration pattern of β-peptide oligomers, and to create alternating heterochiral homo-oligomers by the sequential coupling of (1*S*,2*R*)-*cis*-2-aminocyclopentanecarboxylic acid (ACPC) with (1*R*,2*S*)-*cis*-ACPC, and (1*S*,2*S*)-*trans*-ACPC with (1*R*,2*R*)-*trans*-ACPC.

The controlled self-assembly of foldameric helices leads to helix bundles, vesicle-forming membranes of vertically amphiphilic helices and lyotropic liquid crystals, all of importance for future applications.[7] Conformational order at a higher level (tertiary or quaternary structure) is important not only as a fundamental structural goal, but also to endow foldamers with important functions, such those of as catalyst, which, among proteins, generally require discrete tertiary folding. Self-assembling phenomena have been reported only for H14 helices, and especially for 6-membered side-chain-containing segments. While the H10/12 helix has been stated to be inherently stable, its controlled self-assembly has not been observed. For the β-peptide helices, minor changes in the side-chain topology, size and shape resulted in large effects on the prevailing 3D structure. Our aim was to investigate the existing

6

secondary structure and self-association in the function of different 6-membered alicyclic side-chain shapes with alternating heterochiral homo-oligomers.

The prevailing 3D structure of a foldamer is determined by many factors, such as the residue type, the side-chain topology and chemistry, etc. An extension of the alternating backbone configuration to several residues along the backbone and simultaneous variation of the residue types (α- and β-residues) can lead to novel periodic secondary structures. Through study of the existing secondary structures as a function of the backbone configurations, we planned to establish an intuitive foldamer design tool, a simple stereochemical patterning approach (SPA), with which the prevailing 3D structure can be predicted even for a non-homogeneous back-bone.

2. Literature background

2.1. Foldamers

Foldamers are artificial biomimetic self-organizing polymers.[1-9] Similarly to natural peptides, proteins, RNA and DNA, they fold into specific, well-defined 3D structures, with three properties of crucial importance: (*i*) hierarchical organization (primary, secondary and tertiary structure); (*ii*) cooperativity in folding; and (*iii*) sequence heterogenity. There are two major types of foldamers: (*i*) the aromatic foldamers with aromatic carbon atoms in the back-bone, and (*ii*) aliphatic peptide-like foldamers with saturated carbon atoms between the H-bond pillar moieties.[8] The most prominent representatives of this field are the β- and γ-peptides,[6,10] oligoureas[11] and azapeptides.[12] Chimera sequences with a mixed backbone pattern also exist, such as α/β-[13] and β/γ-peptides[14] with different amino acid compositions (Figure 1).

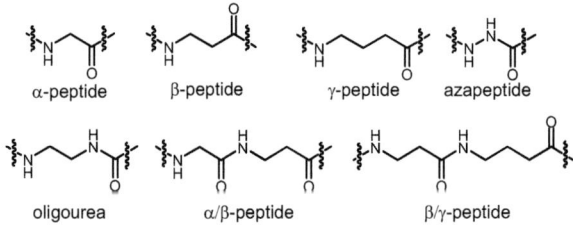

Figure 1. Selected aliphatic foldamer backbones

2.2. β-Peptidic foldamers

2.2.1. Secondary structure

β-Peptides consist of β-amino acids. Interestingly, the first β-amino acids were not synthesized in a laboratory; they were found in Miller's famous experiment,[15] mimicking the conditions of primitive Earth and were also discovered in some extraterrestrial asteroids.[16] In consequence of the additional methylene group, their substitution pattern shows higher diversity (Figure 2).

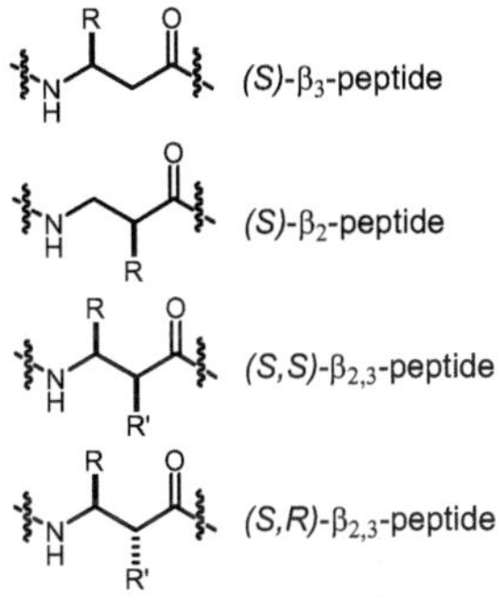

Figure 2. General substitution pattern of β-peptides

The backbone can be mono- or disubstituted. In the latter case, the side-chain can form a cyclic ring. The structural diversity can be increased further by chirality. In cyclic amino acids the relative configuration of the amino and carboxyl functions can be *cis* or *trans*, with a total of 4 possible stereoisomers (Figure 3).

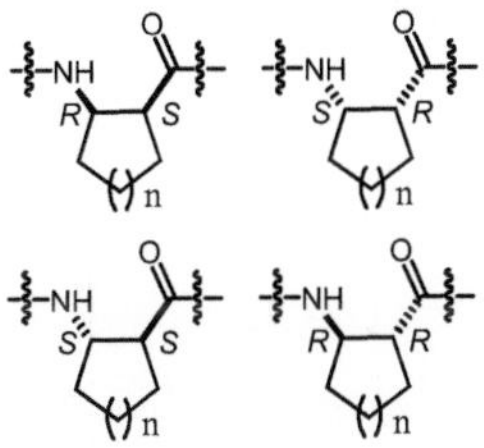

Figure 3. β-Amino acid enantiomers with cyclic side-chains

It might be expected that the insertion of an additional methylene group in the backbone would increase the conformational flexibility of the β-amino acids, resulting in unfavorable entropy associated with secondary structure formation.[17] Fortunately, this is not the case: designed β-peptides fold into specific secondary structures at even shorter chainlengths than those for natural α-peptides.[1,18] A further consequence of the additional methylene group is that it is necessary to introduce an additional dihedral angle to describe the backbone conformation. In the notation of Banerjee and Balaram it is called θ (Figure 4).[19]

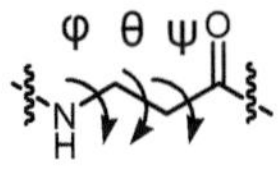

9

Figure 4. Definition of the torsions describing the β-peptide backbone conformation

The secondary structures of β-peptides can be divided into three major groups: helices, strands and turns; the major types are shown in Figure 5.

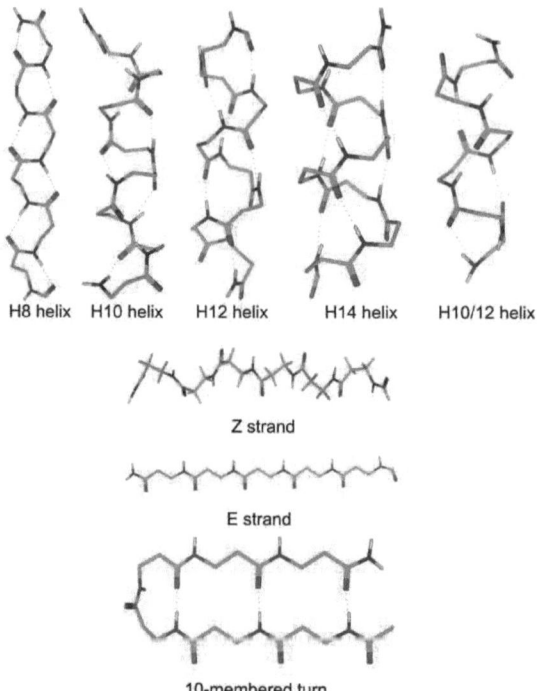

H8 helix H10 helix H12 helix H14 helix H10/12 helix

Z strand

E strand

10-membered turn

Figure 5. Selected examples of experimentally observed β-peptide secondary structures

The most popular nomenclature utilizes the numbers of atoms in one H-bonded pseudoring. This is very useful for helix geometries. The helical secondary structures can have the backbone H-bond orientation from the donor to the acceptor either parallel or antiparallel according to the direction of the helix from the N-terminus to the C-terminus. For the H10 and H14 helices, the orientation is parallel. The H8 and H12 structures are stabilized by antiparallel H-bonds. The very interesting H10/12 helix exists with concatenated H-bonds in alternating orientation; it therefore has negligible dipole moment. A noteworthy property of the β-peptide helices is that the

10

helicity can be clockwise (*M*) or counter-clockwise (*P*); and can be effectively controlled by the backbone stereochemistry.

For strand-like structures, the H-bonded pseudoring-based notation is not useful. The orientation of the peptide bonds can result in two different positions: a uniform and an alternating peptide bond pattern, leading to a polar and a non-polar strand, respectively. The latter backbone conformation displays a zig-zag motif, leading to the name Z strand. The polar strand exhibits an elongated conformation, resulting in the name E strand.[20]

The first described and most widely investigated structure is the H14 helix. This was synthesized by the Gellman group from enantiomerically pure (1*R*,2*R*)-*trans*-2-aminocyclohexanecarboxylic acid (ACHC). They showed that the tetramer, pentamer and hexamer with protecting groups on the N- and C-terminals (Figure 6a) formed the H14 helix in both the solid and the solution phase.[21,22] The Seebach group synthesized homologated β_3-amino acids via Arndt-Eistert homologation, and the peptides also demonstrated H14 geometry.[23-25] The peptides derived from the homologation of natural L-α-amino acids form an *M* helix. The helix is stabilized by an amide proton at position *i* and an (*i*+3) carbonyl group, leading to H-bonding interactions forming a series of intercatenated 14-membered pseudorings. A comparison with the natural α-helix reveals many differences: (*i*) the helix is slightly wider, (*ii*) it has a shorter pitch height, (*iii*) the net dipole caused by the amide bond orientation is the opposite, and (*iv*) one helical turn is made by 3 residues, which orders the residues atop one another along one face of the helix. In organic solvents, the H14 helix is stable as compared to the α-helical conformation of α-peptides. Similarly to natural α-peptides, the conformational polymorphism has been demonstrated. Without the backbone protecting groups (Figure 6b), the *trans*-ACHC oligomers adopt the H10 helix at short chainlengths (tetramer) and the H14 helix appears only at longer sequences.[26]

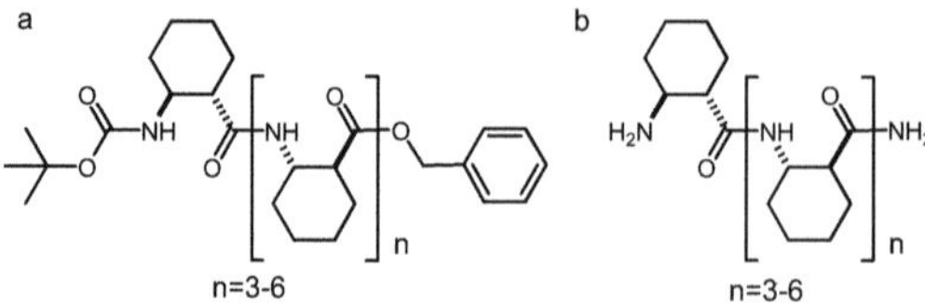

Figure 6. (1*S*,2*S*)-*trans*-ACHC oligomers with (a) and without (b) a backbone protecting group

The H14 helix made from β_3-amino acids appears to be less stable in aqueous medium.[27,28] To overcome this drawback, different methodologies have been devised: (*i*) insertion of conformationally constrained alicyclic (1*S*,2*S*)-*trans*-ACHC residues,[29] or (*ii*) the use of side-chain interactions e.g. a metathesis reaction,[30] disulfide bridges[31] or saltbridges. [32-34]

Immediately after the discovery of the H14 helix, a search began for new helical structures to expand the conformational space of β-peptide helices. It was found that the oligomers of *trans*-ACPC fold into the H12 helix.[35-38] However, these structures were not soluble in aqueous medium.[1] To overcome these difficulties, *trans*-4-aminopyrrolidine-3-carboxylic acid (APC)-containing oligomers were synthesized to furnish water-soluble and water-resistant H12 helices (Figure 7).[39]

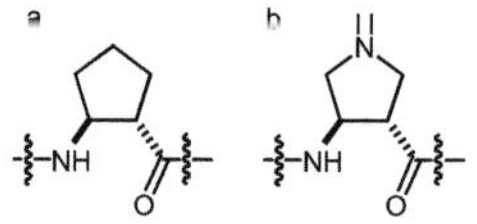

Figure 7. (1*S*,2*S*)-*trans*-ACPC (a) and (3*S*,4*R*)-*trans*-APC (b), which forms the H12 helix

Proteinogenic side-chains have additionally been introduced through the *trans*-APC residues by sulfonylation of the amine group.[40] Investigation of the tolerance of acyclic residues in the H12 helix has shown that the incorporation of β_3- or β_2-amino acids slightly destabilizes the structure, but this effect can be repressed by the incorporation of 5-membered cyclic side-chains.[41,42]

One helical turn in the H12 helix is made by ~ 2.5 residues and displays some similarities to the natural α-helix: *(i)* the structure is stabilized by intercatenated *i* – (*i* + 3) H-bonds, *(ii)* the helix polarity and amide bond direction are the same, and *(iii)*

12

the helix diameter is 2.3 Å. In view of the structural connections to the α-helix, the H12 helix was postulated as the ideal α-helix mimetic foldamer, and an extensive search therefore began for novel methods to stabilize the H12 structures. It was shown that, due to side-chain steric repulsion in organic solvents, ethanoanthracene-based oligomers[43] and apopinane-based peptides[44] form stable H12 helices. Water-resistant H12 helices have been obtained by the application of aza-ACPC.[45]

The alternate coupling of β_2- and β_3-monosubstituted β-amino acids led to a H10/12 helix conformation.[24,25] The structure is stabilized by concatenated 10 and 12-membered H-bonded pseudorings. The amide bonds exhibit alternating orientation along the backbone, leading to a negligible net dipole as compared to other helices. The same helix geometry has been described with C-linked carbohydrate side-chains and an alternating backbone configuration, extending the variety of side-chain functionalities.[10] A H10/12 helix has been designed by using β-alanine and a constrained cyclic β-amino acid with a carbohydrate-derived 5-membered ring.[46] This helical structure with left-handed helicity has also been constructed by using another sugar derivative β-amino acid oligomer.[47]

In parallel with the characterization of the helical structures, strands and pleated sheet structures have been described. Firstly, only hairpin-like structures allowed the construction of sheet-like structures, where two strands were linked together with a turn motif.[48-50] Hairpin-like pleated sheets have been created by α/β-disubstituted sequences connected together with β_2/β_3-dipeptides.[51,52] The use of alicyclic β-amino acids meant a significant development in this field. The oligomers constructed from (1R,2S)-ACPC form a Z6 strand without the need for any extra stabilizing turn motif between the two strands (Figure 8).[53]

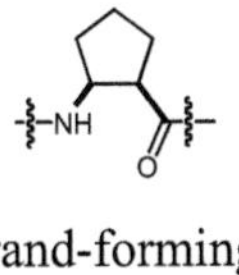

Figure 8. The Z6 strand-forming (1R,2S)-cis-ACPC

The easiest way to increase the chemical diversity of the side-chain functionalities is the insertion of α-amino acids into the β-peptide chain, leading to α/β-chimera

13

sequences. The first systematically studied α/β-peptide was composed of L-alanine and (1S,2S,3S)-2-amino-3-(methoxycarbonyl)cyclopropanecarboxylic acid residues, leading to an H13 helix.[54] L-α-Amino acids have been inserted between (1S,2S)-cis-ACPC residues. This oligomer had a "split personality", with two rapidly interconverting conformations, e.g. an H11-helix and an H14/15 helix.[55] Homologated β$_3$-amino acids have been coupled with D-α-amino acids to give an H9/11 helix. This structure exhibits an alternating amide bond orientation, similarly to the H10/12 helix.[56,57] "Split personality"-type chimera structures have also been described by the 1:1 combination of a carbohydrate derivative β-amino acid and L-alanine.[58] Both the 2:1 and 1:2 α/β backbone patterns promote helix formation. For the 2:1 α/β peptide the helix is stabilized by 10/11/11 H-bonding pseudorings. The secondary structure formed by the 1:2 α/β chimera sequence is an H11/11/12 helix.[59]

2.2.2. Secondary structure design

The described secondary structures can be tuned efficiently via the side-chains. For open-chain β-amino acids, the effect of substitution on the preferred backbone geometry has been investigated on model systems. For β$_3$-substitution, the model compound is N-isopropyl formamide and the local effect of substitution on φ has been investigated. For β$_2$-amino acids, the effect of the side-chain on ψ has been monitored on isobutyramide as model compound.[60-62] The potential energy curve (PEC) obtained for φ indicates that the preferred value is between 60° and 180° and there is also a sharp minimum with a relatively high barrier at -60°. The PEC for ψ also has a relatively flat minimum between 60° and 180° and a narrow minimum at -60°.[3]

These results are those for the S configuration; the values for the R configuration can easily be obtained by multiplication by -1. A comparison of these results with the dihedral angles of the known helices reveals that a β$_3$- or β$_2$-monosubstitution generally strongly promotes the formation of a helical conformation. Consequently, the first periodic secondary structure obtained for β$_3$-homologated amino acid oligomers was the H14 helix. θ is an important factor in structure formation. For helical structures it should have a gauche conformation. For cyclic residues with (S)-β₂-(S)-β₃ or (R)-β₂-(R)-β₃, φ and ψ can be found in the optimal range and θ is fixed in the gauche conformation. In consequence, β-peptides with cyclic side-chains are known to form helical structures with the shortest chainlengths. The fine-tuning of θ by different cyclic side-chains leads to diverse secondary structures. For *trans*-ACHC oligomers, θ is ~ 60° (real gauche conformation) and gives the expected H14 helix (Figure 9a). For *trans*-ACPC oligomers, θ is ~ 90°, which is intermediate between the gauche and eclipsed conformations; this value leads to a H12 helix (Figure 9b). θ can be efficiently tuned via β₂,₃-disubstitution of the backbone. For the substitution pattern of (R)-β₂-(S)-β₃ or (S)-β₂-(R)-β₃, the antiperiplanar orientation is favored and gives θ ~ 180° (Figure 9c). This geometry is necessary for the polar (elongated) strand and can not be stabilized by H-bonds; it therefore needs to be stabilized by turn-forming segments, where two polar strands are coupled together to give a hairpin-like structure.[51,52] If these two substituents are closed into a ring, the antiperiplanar conformation can not be adopted and formation of the E strand is not possible. Indeed, a zig-zag (Z strand) structure is established by the homooligomers of (1R,2S)-*cis*-ACPC, which is stabilized by 6-membered H-bonded pseudorings and no extra stabilization is needed.[53]

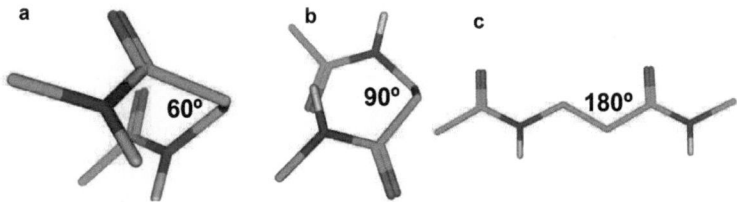

Figure 9. θ for the H14 (a), H12 (b) and E strand (c)

A noteworthy feature is that the flexibility of the residue and the chain length can influence the secondary structure; consequently the *trans*-ACHC tetramer forms a H10 helix with the conformational polymorphism of β-peptides. The nucleation effect of the cyclic residues has also been investigated. The polymers of the β₃-homologated amino acids form the well-known H14 helix, while incorporation of the H12 helix-

15

forming *trans*-ACPC residues into β_3-peptides promotes or nucleates formation of the H12 helix.[41,42]

Long-range side-chain interactions are very important in helix stabilization. In the case of the H14 helix, all the β_2- and β_3-substituents at positions $i - (i + 3)$, and for H12 all the $(i)\beta_3 - (i + 2)\beta_2$ and the $(i)\beta_2 - (i + 3)\beta_3$ substituents are adjacent. Consequently, extra stabilization can be gained by hydrophobic, van der Waals and electrostatic interactions. As described previously, the H14 helix has been efficiently stabilized in aqueous medium by side-chain interactions.[32-34] The secondary-structure H10 and H10/12 helices exclude such interactions.

2.2.3. Stereochemical control over secondary structures

Stereochemistry is an excellent tool with which to tune secondary structures. Natural peptides are mainly composed of L-α-amino acids. However, there are three important exceptions: (*i*) anthrax polypeptide, (*ii*) gramicidin β-helix and (*iii*) the α-sheet. The anthrax polypeptide consists of D-glutamic acid and the amino acids form a γ-peptidic back-bone.[63] These structural differences are the basis of the high virulence and resistance of *Bacillus anthracis* against proteolysis and many antibacterial agents. These structural properties fulfill the definition of foldamers, and consequently the first foldameric structures were created by the Nature. Based on the knowledge gained from Nature, many D-peptidic sequences have been made and used in drug research.[64] The β-helix is formed by the alternating coupling of L- and D-α-amino acids. This structure found in *Bacillus brevis* forms ion channels which damage the attached cell via the same mechanism as for helical amphiphiles.[65] The alternating coupling of L- and D-α-amino acids can also lead to the formation of an α-sheet secondary structure.[66] As there are only 2 enantiomers for α-amino acids, whereas β-amino acids potentially have 4 stereoisomers, higher diversity is foreseen. Consequently, stereochemistry is an excellent tool for enhancing the structural diversity of 3D structures of β-peptide foldamers. The peptides derived from the Arndt-Eistert homologation of natural L-α-amino acids form an H14$_M$ helix, while the D configuration leads to *P* helices. Similarly, the oligomers of (1*S*,2*S*)-*trans*-ACHC

16

form an H14$_M$ helix and the (1R,2R)-$trans$-ACHC forms an H14$_P$ helix. For helix formation, the homologated L-amino acids can be combined only with (1S,2S)-$trans$-ACHC in homochiral systems. The alicyclic amino acids with $trans$ relative configuration and 5 or 6-membered side-chains are known as helix-forming building blocks. On change of the stereochemistry to the cis relative configuration, the prevailing secondary structure will be strand-like.[53] A β-peptide chain composed of alternating L-β$_3$- and (S)-β$_2$-amino acids folds preferentially into an H10/12$_P$ helix. The stereochemically appropriate combination of α- and β-amino acids raises a very important question. The results have revealed that only merging of L-α-amino acids and (1S,2S)-$trans$-ACPC gives a well-defined helical secondary structure in homochiral systems,[55] which is dependent on the α/β-amino acid ratio.[59] Similarly, the combination of L-α-amino acids with (1S,2S,3S)-2-amino-3-(methoxycarbonyl)cyclopropanecarboxylic acid residues affords a helical structure. Consequently, the sugar derivative β-amino acid with R absolute configuration can be combined with D-alanine, leading to helices.[58] Similarly to the alternating H10/12 helix described by Seebach et al., an H9/10 helix has been created by the coupling of D-α- and L-homologated-β-amino acids.[56,57]

2.2.4. Tertiary structure motifs

Foldamers have been defined with the property of hierarchical self-organization. The search for tertiary structure motifs of foldamers is a great challenge, because mimicking the functions of enzymes, transport proteins, structural proteins, motor proteins and receptors requires a tertiary or quaternary structure for the function. An early observation of a helix bundle structure was performed by analytical ultracentrifugation of an amphiphilic H14 helical decamer sequence.[67] Different approaches have been developed for the stabilization of helix bundles. Nucleobase functionalized H14 helices have been designed where a reversible self-association has been gained, driven by the H-bonding between the base-pairs.[68] Zinc binding motifs have also been incorporated into β-peptides, leading to dimer formation through complexation of the zincion. They behave similarly to the zinc-finger-type transcription factors.[69] Disulfide-bridged H14 helices have also been described where the association provided extra stabilization for the secondary structure.[70] Others have focused on the solvent-driven self-association of amphiphilic β-peptides. The hydrophobic packing of the helices resulted in tetrameric[71] or octameric[72] bundles. Such self-assembly has been described for α/β-chimera sequences, and side-chain-dependent trimer or tetramer structures have been reported.[73,74] The results were obtained in the solid phase and the structures were investigated by X-ray crystallography.

A similar self-assembly has been observed in the appearance of birefingent liquid crystals for H14 helix-forming β-peptides.[75,76] Helix-forming (1S,2S)-trans-ACHC oligomers with vertical amphiphilicity were shown to form helix bundles in the solution phase by NMR investigations and dynamic light scattering (DLS) measurements. Interestingly, the transmission electron microscopy (TEM) images of these oligomers revealed the formation of multilamellar vesicles as evidence of self-assembly.[77]

On the other hand, the Z6 strand-forming (1R,2S)-cis-ACPC oligomers lead to amyloid-like fibrils. For the heptamer (1R,2S)-cis-ACPC, the height of the fibril after 1 week of incubation corresponded to the length of the β-peptide. The width was 30-40 nm, which can be formed by the self-association of 30-40 monomers, and the length of the associate was in the μm range.[77]

2.2.5. Pharmaceutical applications

The significance of β-peptide foldamers in drug discovery is warranted by two major facts: (*i*) a stable and predictable secondary or tertiary structure, and (*ii*) resistance against enzymatic degradation *in vitro* and *in vivo*.[78,79] Natural antimicrobial peptides are important parts of the innate immune system. They kill bacteria by lowering the integrity of the membranes of their targets.[80,81] These peptides are helical amphiphiles, positioning the polar and nonpolar side-chains on two different surfaces of the helix. The insertion of such a structure leads to disruption of the bacterial membrane. They can cause harm by enhancing lability in

18

the surface pressure and chemical potential of the bilayer, and forming ion channels. These two pathways can lead to generalized disruption of the membrane, with collapse of the transmembrane potential, leakage of intracellular contents and the death of the bacterium as the final result.[82] On the basis of the structures of the antibacterial α-peptides, β-peptides with helical amphiphilicity were designed.[83-85] The first generation of these oligomers were potent antimicrobial agents, but they also exerted significant hemolytic activity, damaging human erythrocytes. This effect is caused by the hydrophobic character of the compounds. A significant improvement in this field resulted from the incorporation of positively charged cyclic residues into a helical structure, leading to high antibacterial and minimal hemolytic activity. The insertion of positive charges leads to higher affinity for the negatively charged bacterial membranes and lower affinity for the hydrophobic erythrocytes.[84] Interestingly, some natural host-defense peptides adopt an α-helical conformation only in the bacterial cell membrane, leading to separation of the hydrophobic and hydrophilic side-chains, giving an amphiphilic structure.[86] From α/β-chimera sequences, peptides without predefined secondary structure have been formed by a random construction of lipophilic and hydrophilic units. These oligomers exhibited significant activity against bacteria with low hemolytic potential.[87,88]

The major aim in foldamer research has always been the modification of protein-protein interactions (PPIs). Many proteins do not have a distinct binding pocket but only flat surfaces. Consequently, they are undruggable with small molecules.[89] In such cases, foldamers could be very useful compounds, offering a higher interaction surface. The first application in this field was the inhibition of fat and cholesterol absorption in the small intestine by antagonizing the function of the SR-B1 lipid transport protein by a β-nonapeptide helix.[90] This compound was designed on the basis of a known α-peptide inhibitor.

Polycationic oligomers have been synthesized with the aim of binding to a membrane or oligonucleotide. The transcription of HIV RNA requires interaction of the Tat protein with a hairpin-like RNA segment called the transcriptional activator-responsive element (TAR). A Tat analog TAR-binding β-peptide with positively charged side-chains was designed with nanomolar affinity.[91] Polycationic peptides are known to penetrate cell membranes. β_3-Homoarginine-containing sequences have been described which allow efficient membrane transport in a chain length-dependent manner. These peptides enter the cell in some cases in an endocytosis-dependent process, while in other cases they cross the cell membrane in a different but still unknown process even at 4°C and in the presence of NaN_3.[6,92,93] Homolysine-rich sequences have also been described and used as gene delivery agents. The results indicated effective membrane transport and interaction with DNA.[94]

A significant development was achieved in the biomedical applications of β-peptides through inhibition of the PPIs between the tumor suppressor protein p53 and the oncoprotein hDM2 by a *de novo* designed sequence.[95,96] The p53 protein plays a crucial role in apoptosis and hDM2 negatively regulates its function. Disruption of this interaction is a major aim in cancer research.[97] The activation domain of p53 recognizes hDM2 through an α-helix. β-Peptides which mimic this helix have been designed, and the cocrystallization with hDM2 demonstrated binding with similar geometry to the α-helix. PPIs between the two domains of the AIDS-related HIVgp41 protein were blocked with an epitope mimic β_3-decapeptide.[98,99] Chimera sequences made by the coupling of α- and β-amino acids have also gained applications through the modification of PPIs. Inhibition of the HIVgp41 protein by α/β-chimera sequences has also been described.[100] The overexpression of antiapoptotic protein Bcl-x_L is detectable in many tumorous cells. Its function is regulated by the proapoptotic Bak or Bad factors and they bind to Bcl-x_L via a BH3 domain. The structures of BH3 have led to the design of β- and α/β-peptides as antagonists for Bcl-x_L. One oligomer showed a very high inhibitory potential, with $K_I \sim 1$ nM.

For an H12 helical *trans*-ACPC-based foldamer, specific gamma-secretase inhibition with nanomolar activity has been described. These β-peptides can be lead compounds in the field of Alzheimer disease treatment research.[101]

Not only helical structures have been used as skeletons for the formation of bioactive foldamers. Somatostatin has a hairpin-like 3D conformation, and structurally analogous open-chain β-tetrapeptides with turn-like secondary structures and cyclic oligomers have been designed. These (*i*) are high-affinity agonists for somatostatin sst4-receptor, (*ii*) display good oral bioavailability, (*iii*) are resistant to biodegradation and (*iv*) are excreted within 4 days.[102,103] Furthermore, cyclic β-tripeptides and their derivatives have been used as antagonists for a tumor necrosis factor receptor superfamily member CD40 receptor. The ligands interact very efficiently (K_D = 2.4 nM) and induce apoptosis in lymphoma and leukemia cells.[104,105]

The ADME properties of structurally diverse β-peptides have also been investigated. The radioactively labeled compounds generally demonstrated low oral bioavailability, outstanding proteolytic and metabolic resistance and highly structure-dependent distribution and elimination in rats.[103,106,107]

2.3. Synthesis of β-peptides

The synthesis of β-peptides can be performed in the following ways: (*i*) solution-phase synthesis, (*ii*) synthesis on a solid support either with a *tert*-butoxycarbonyl/benzyl (Boc/Bzl)[108] or with a 9-fluorenylmethoxycarbonyl/*tert*-butyl (Fmoc/'Bu) technique.[109]

For homo-oligomeric sequences, the fragment condensation method in the solution phase offers advantages. This method was used for the first synthesis of *trans*-ACHC homo-oligomers by applying Boc for amino protection and the benzyl ester for carboxyl protection. The activation was performed with 1-ethyl-3-(3-dimethylaminopropyl)carbodiimide (EDCI) and 4-(N,N-dimethylamino)pyridine (DMAP).[21]

A similar solution-phase strategy has been used for the synthesis of β-peptides from β₃-homologated amino acids.[23,24] This method suffers from various drawbacks however: the reaction time is very long, and the carbodiimide activation can lead to racemization. The decreased solubility caused by the protected oligomer can give truncated sequences, resulting in difficulties in product purification. An additional consequence of the presence of protecting groups on the termini is their effects on the evolving secondary structure.[26]

21

To avoid the mentioned difficulties and the need for more complex peptides, solid-phase peptide synthesis (SPPS) has been applied,[24-26,31,32,53] with both the Boc-based and the Fmoc-based strategies. The coupling of conformationally constrained alicyclic β-amino acids is often difficult and becomes very complicated in homo-oligomers after the fourth residue. Sharpening the problem, the visualization of the incomplete coupling can not be performed perfectly with the Kaiser test.[110] Only the test cleavage of aliquots followed by HPLC-MS measurement can yield exact information. To avoid the formation of truncated sequences, novel coupling agents such as the *in situ* amino acid fluoride-generating fluoro-N,N,N',N'-tetramethylformamidinium hexafluorophosphate (TFFH) have been applied effectively.[26] The use of microwave irradiation, 1-methyl-2-pyrrolidinone (NMP) as solvent and chaotropic salts such as LiCl and the application of an active ester-forming uronium-type reagent such as O-(benzotriazol-1-yl)-N,N,N',N'-tetramethyluronium hexafluorophosphate (HBTU) led to increased coupling efficiency and a reduced reaction time. With this methodology, one coupling and deprotection step can be performed within 45 min.[111] The use of Fmoc methodology is favorable relative to the Boc methodology, because it utilizes milder conditions. However, the deprotection step for the β-peptides can be problematic and can lead to incomplete reactions. The use of a longer reaction time or the application of stronger bases such as 1,8-diazabicyclo[5.4.0]undec-7-ene (DBU) did not improve the result. Only increased temperature (60 °C) led to a complete reaction.[111]

The incorporation of trifunctional homologated β-amino acids has also been performed on a solid support with either Boc or Fmoc techniques.[18,23,25] The Boc methodology with 4-methylbenzhydrylamine (MBHA) resin[112] furnished peptide amides,[26] while the Fmoc technique with either Rink amide[73,113] or Wang resin[103,114] gave peptide amides or acids, respectively. The side-chain protecting groups used in the SPPS of α-peptides have been efficiently used in the case of β-peptides too. The cleavage with MBHA resin was performed with HF in the presence of anisole and thioanisole, while with Rink amide resin it was acchieved with a mixture of trifluoroacetic acid (TFA), 1,2-ethanedithiol (EDT), triisopropylsilane (TIS) and water (Scheme 1).[115] The peptide purification was performed under standard conditions by means of reversed-phase high-performance liquid chromatography (RP-HPLC).[116]

For longer sequences, the thioligation method has been used effectively.[117] The thioester was synthesized by the method of Ingentio et al.,[118] based on the acylsulfonamide safety catch linker.[119] The displacement reaction was performed overnight at elevated temperature (80 °C). After cleavage of the peptides, the native chemical ligation was performed under standard conditions in aqueous solution.[117]

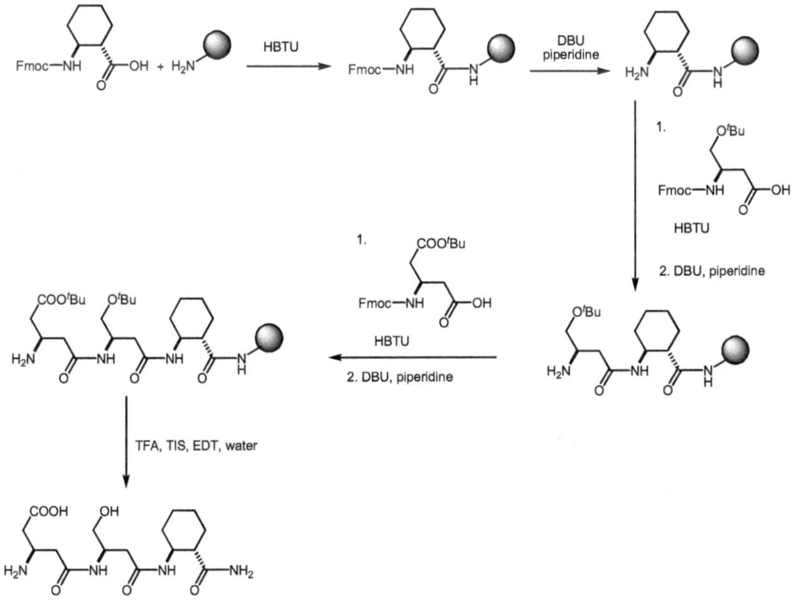

Scheme 1. Fmoc-based SPPS with various amino acids on Rink amide resin

3. Methods

3.1. Synthesis

Peptide Synthesis. Peptides were synthesized on a solid support by either the Boc or the Fmoc technique. The Boc methodology was used on MBHA resin[112] (0.63 mmol g^{-1}), and the syntheses were carried out manually on a 0.25 mmol scale. Couplings were performed with 3 equivalents of dicyclohexylcarbodiimide (DCC) and hydroxybenzotriazole (HOBT), generally without difficulties in dichloromethane (DCM) as solvent. The reaction time was 3 h. In the event of difficult coupling steps, the amount of truncated sequences increased, which could not be avoided through the use of increased equivalents of the amino acid, more active coupling agents or a longer reaction time. After each coupling step the resin was washed 3 times with DCM, once with MeOH and 3 times with DCM. The amino acid incorporation was monitored by means of the ninhydrin test[110] and occasionally by the cleavage of aliquots from the resin. The Boc protecting group was removed by using 50% TFA solution in DCM in two steps with reaction times of 20 and 15 min. After the deprotection step, the resin was washed with the same methodology. The resin was neutralized with 10% triethylamine solution in DCM. The peptide sequences were cleaved from the resin with liquid HF/dimethyl sulfide/p-cresol/p-thiocresol (86:6:4:2, v/v) at 0 °C for 1 h. The HF was removed, and the resulting free peptides were precipitated with cooled dried diethyl ether. The precipitated peptides were filtered, washed, solubilized in 10% aqueous acetic acid and lyophilized. Crude peptides were purified by using RP-HPLC, with a Nucleosil C18 7 µm 100 Å column (16 mm x 250 mm).[120] The HPLC apparatus was made by Knauer.[121] The solvent system was as follows: 0.1% TFA in water; 0.1% TFA, 80% acetonitrile in water; a linear gradient was used during 60 min, at a flow rate of 3.5 mL min^{-1}, with detection at 206 nm. The purity of the fractions was determined by analytical RP-HPLC, using an Agilent 1100 HPLC system[122] equipped with a Phenomenex Luna C18 100 Å 5 µm column (4.6 mm x 250 mm)[123] and the pure fractions were pooled and lyophilized. The purified peptides were characterized by mass spectrometry (MS), using a Finnigan TSQ 7000 tandem quadrupole mass spectrometer equipped with an electrospray ion source.[124]

With Fmoc chemistry, the peptide chain was elongated on TentaGel R RAM resin (0.19 mmol g^{-1})[125] with a Rink amide linker on a 0.1 mmol scale manually. The coupling was performed in two steps. In the first step, 3 equivalents of Fmoc-protected amino acid, 3 equivalents of the uronium coupling agent O-(7-azabenzotriazol-1-yl)-N,N,N′,N′-tetramethyluronium hexafluorophosphate (HATU)[126] and 6 equivalents of N,N-diisopropylethylamine (DIPEA) were used in N,N-dimethylformamide (DMF) as solvent with shaking for 3 h. The second coupling was performed with 1 equivalent of amino acid, 1 equivalent of HATU and 2 equivalents of DIPEA. After the coupling steps, the resin was washed 3 times with DMF, once with MeOH and 3 times with DCM. No truncated sequences were observed under these coupling conditions. Deprotection was performed with 2% DBU and 2% piperidine in DMF in two steps, with reaction times of 5 and 15 min. The resin was washed with the same solvents as described previously. The cleavage was performed with TFA/water/DL-dithiothreitol (DTT)/TIS (90:5:2.5:2.5) at 0 °C for 1 h. The purification was carried out by RP-HPLC, using a Phenomenex Luna C18 100 Å 10 µm column (10 mm x 250 mm).[123] The HPLC apparatus was made by JASCO.[127] The solvent system used was as follows: 0.1% TFA in water; 0.1% TFA in 80% acetonitrile in water; a linear gradient was used during 60 min, at a flow rate of 4.0 mL min^{-1}, with detection at 206 nm. The purities of the fractions were determined by analytical RP-HPLC using a JASCO HPLC system[127] with a Phenomenex Luna C18 100 Å 5 µm column (4.6 mm x 250 mm)[123] and the pure fractions were pooled and lyophilized. The purified peptides were characterized by MS, using a Finnigan MAT 95S sector field mass spectrometer equipped with an electrospray ion source at a resolution of ~ 1000 - 1500.[124]

Flow reactor hydrogenation: The reaction was performed in an H-Cube® flow reactor apparatus.[128] The previously purified unsaturated precursor was dissolved in MeOH and the hydrogenation was carried out on 5% Pd/charcoal catalyst, at 50 atm, with a flow rate of 1 mL min^{-1} and 5 recirculations. The characterization was performed by using MS with a Finnigan MAT 95S sector field mass spectrometer equipped with an electrospray ion source at a resolution of ~ 1000 - 1500[124] and the purity was determined by RP-HPLC analysis.[127]

3.2. Structure investigations

Nuclear magnetic resonance (NMR) measurements were performed on Bruker Avance DRX 400 and 600 MHz NMR spectrometers[129] with the deuterium signal of the solvent as the lock. The peptide oligomers were investigated in 4 or 8 mM solution in CD_3OD, CD_3OH, DMSO-d_6 and water (H_2O/D_2O 9:1). The signal assignment was performed by the application of 2D NMR measurements, such as TOCSY (total correlation spectroscopy),[130,131] ROESY (rotating frame Overhauser effect spectroscpoy)[132] and NOESY (nuclear Overhauser effect spectroscopy).[133,134] During TOCSY measurements, homonuclear coherent magnetization transfer is performed and cross-peaks are generated between all members of a coupled spin system. The MLEV 17 mixing sequence was used with a mixing time of 80 ms; 32 scans, 2k time domain points and 512 increments. During ROESY measurements, homonuclear incoherent magnetization transfer is made. The experiment allows the correlation of nuclei through space where the distance is smaller than 5 Å. For medium-sized molecules (~1000 Da), ROESY measurements can be used.[135] For the ROESY spinlock, mixing times of 225 ms and 400 ms were used; the number of scans was 64, and 2k time domain points and 512 increments were applied. The TOCSY and ROESY measurements were performed at 303.1 K, 296.1 K, 277.1 K and 273.1 K. For processing, a cosine bell function was applied before transformation. The ROESY cross-peak volumes were integrated by using the Topspin 2.0 software.[136] The proton distances were calculated from the ROESY cross-peak volumes by using the isolated spin-pair approximation, with a distance between vicinal α- and β-protons of 2.4 Å as reference. The integrated volumes were offset-compensated for the ROESY off-resonance effects.

Electronic circular dichroism (ECD) measurements were performed on a JASCO J810 dichrograph[137] at 25 °C in a 0.02 cm cell. Eight spectra were measured

26

for each sample and the baseline spectrum recorded with only the solvent was subtracted from the averaged data. The concentration of the sample solutions was 1 mM in CD_3OH and H_2O. Molar circular dichroism ($\Delta\varepsilon$) is given in M^{-1} cm^{-1} and the data were normalized for the number of chromophores. For spectrum interpretation, the Spectra Manager 2.0[138] software was used.

Vibrational circular dichroism (VCD) spectra at a resolution of 4 cm^{-1} were recorded in DCM and DMSO-d_6 solutions with a Bruker PMA 37 VCD/PM-IRRAS module connected to an Equinox 55 FTIR spectrometer. The ZnSe photoelastic modulator of the instrument was set to 1600 cm^{-1} and an optical filter with a transmission range of 1960-1250 cm^{-1} was used in order to increase the sensitivity in the carbonyl region. The instrument was calibrated for VCD intensity with a CdS multiple-wave plate. A CaF_2 cell with a pathlength of 0.207 mm and a sample concentration of 10 mg/ml were used. VCD spectra were obtained as averages of 21,000 scans, corresponding to a measurement time of 6 h. Baseline correction was achieved by subtracting the spectrum of the solvent obtained under the same conditions. VCD has become a powerful technique for studies of the stereochemical or conformational properties of molecules. VCD spectra have a high structural information content and, in a comparison of the *ab initio* calculated VCD spectra at the B3LYP/6-311 G** level of theory with the experimental one, the structural hypothesis can be evaluated directly.

Molecular mechanical simulations were carried out in the Chemical Computing Group's Molecular Operating Environment (MOE) software.[139] The MMFF94x[140] force field was used for the energy calculations without a 15 Å cut-off for van der Waals and Coulomb interactions, and the distance-dependent dielectric constant (εr) was set to $\varepsilon=1.8$ (corresponding to MeOH). The conformational sampling was performed by using the restrained simulated annealing (RSA) or the hybrid Monte Carlo (MC) /molecular dynamics (MD) simulation. Before RSA, a random structure set of 100 molecules was generated by saving the conformations during a 100 ps dynamics simulation at 1000 K every 1000 steps. The RSA was performed for each structure with an exponential temperature profile in 75 steps, and a total duration of

25 ps, and the H-bond restraints were applied as a 10 kcal mol^{-1} Å^{-2} penalty function. Minimization was applied after every RSA in a cascade manner, using the steepest-descent, conjugate gradient and truncated Newton algorithm.

The MC-MD was run at 300 K with a random MC sampling step after every 10 MD steps, with a step size of 2 fs for 20 ns, and the conformations were saved after every 1000 MD steps, which resulted in 10000 structures. The NMR distance restraints were applied as a flat-bottomed quadratic penalty term with a force constant of 5 kcal mol^{-1} Å^{-2}. The final conformations were minimized to a gradient of 0.05 kcal mol^{-1} and the minimization was applied in a cascade manner, using the steepest-descent, conjugate gradient and truncated Newton algorithm. The computations were carried out on a HP xw6000 workstation.

Ab initio **calculations** were performed by using the Gaussian03[141] software. The molecular structure, stereochemistry and geometry were exclusively defined in terms of their z-matrix internal coordinate system. The optimizations were carried out at the RHF/3-21G and B3LYP/6-311G** levels of theory with a default set-up. For the solvent calculations, the polarizable continuum model (PCM)[142] was used. The calculations were performed on the computational resources (Fujitsu-Siemens and SGI cluster) of the High-Performance Computing Group (HPC) at the University of Szeged.

3.3 Particle size measurement

Dynamic Light Scattering (DLS)[143] measurements can provide information about the structural and dynamic properties of matter. When a beam of light passes through a colloidal system, the particles scatter some of the light. The peptides were dissolved in MilliQ water to a concentration of 4.0 mM. The solution was sonicated for 5 min and filtered through a sterile syringe filter equipped with a PVDF membrane (pore size: 100 nm, Millipore, Billerica, MA, USA). DLS measurements were performed at 25 °C on a Malvern Zetasizer Nano ZS Instrument (Malvern Instruments Ltd., Worcestershire, UK) equipped with a He-Ne laser (633 nm), by means of Non-Invasive Back Scatter (NIBS®) technology, which means detection of the scattered

28

light at an angle of 173°. The translational diffusion coefficients were obtained from the measured autocorrelation functions by using the regularization algorithm CONTIN built into the software package Dispersion Technology Software 5.1 (Malvern Instruments Ltd., Worcestershire, UK). The correlation function and distribution of the apparent hydrodynamic diameter (d_h) over the scattered intensity of the samples were determined on the basis of 3×14 scans.

Transmission electron microscopy (TEM)[144a] can give information about the morphology of the examined object formed in a colloidal system. The peptides were dissolved in MilliQ water to a concentration of 4.0 mM. The solution was sonicated for 5 min and filtered through a sterile syringe filter equipped with a PVDF membrane (pore size: 100 nm, Millipore, Billerica, MA, USA). Droplets of 10 μL of solutions were placed onto the specimen holder, a carbon-film-coated 400 mesh copper grid (Electron Microscopy Sciences, Washington DC). First, the solution of the aggregated sample was applied to the grid and incubated for 2 min, while the particles accessed the specimen support by Brownian motion and adhered to it. The specimens were fixed with 0.5% (v/v) glutaraldehyde solution (for 1 min), washed three times with MilliQ water, and finally stained with 2% (w/v) uranyl acetate. Specimens were studied with a Philips CM 10 transmission electron microscope (FEI Company, Hillsboro, Oregon) operating at 100 kV. Images were taken with a Megaview II Soft Imaging System, routinely at magnifications of ×25000, ×46000 and ×64000, and analyzed with an AnalySis 3.2 software package (Soft Imaging System GmbH, Münster, Germany).

The particle size measurements were performed with the cooperation of Dr. Lívia Fülöp from the Department of Medical Chemistry University of Szeged.

4. Results and discussion

4.1. Fine-tuning of β-peptide self-organization by alternating backbone stereochemistry

For assessment of the potential of the alternating heterochiral approach, we chose ACPC monomers, because all 4 enantiomers are available synthetically in a reasonably convenient way for this β-residue,[145] and the reference secondary structures formed by the homochiral sequences are known, e.g. the homochiral homo-oligomers of (1S,2S)-*trans*-ACPC are known to facilitate an H12 helix, while (1R,2S)-*cis*-ACPC residues promote the Z6 strand conformation. Both alternating heterochiral *cis*-ACPC (H-[(1R,2R)-ACPC-(1R,2R)-ACPC]$_n$-NH$_2$) and alternating heterochiral *trans*-ACPC (H-[(1R,2R)-ACPC-(1S,2S)-ACPC]$_m$-NH$_2$) oligomers were constructed (Scheme 2).

Scheme 2. The studied alternating heterochiral β-peptide sequences. Reproduced with permission from ref. I.

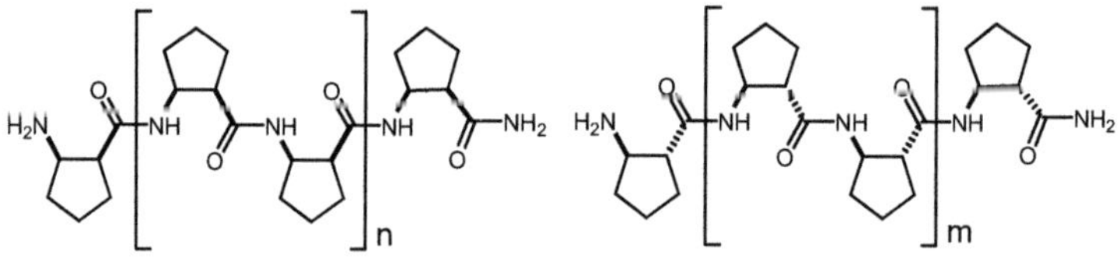

1, n=1; 2, n=2   **3, m=1; 4, m=2**

First, molecular modeling was carried out for hexamers **2** and **4**. A conformational search was performed by molecular mechanics calculation, followed by *ab initio* quantum chemical calculations. An RSA computation was carried out by using molecular mechanics with the MMFF94x force field without any distance restraint. The resulting conformational pool was analyzed, the conformers from the lowest-energy conformational families of **2** and **4** were chosen and further optimizations were performed. The HF/3-21G level of theory in vacuum was utilized first, for the *ab initio* quantum chemical geometry optimizations. The structures converged to the corresponding local minimum of the potential energy surface. Further optimizations were performed in order to take into account the effects of larger basis sets and the electron correlation at the B3LYP/6-311G** level (Figure 10). For **2**, both the RSA and the *ab initio* optimization predicted the H10/12 helix as the prevailing conformer, while for **4** the H8 strand-like conformation was the most stable.

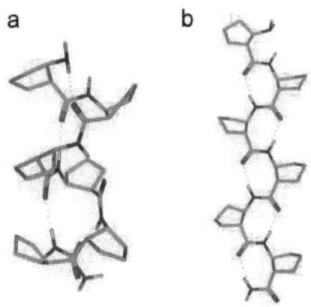

Figure 10. *Ab initio* geometries of **2** (a) and **4** (b), calculated in vacuum at the B3LYP/6-311G** level of theory. Reproduced with permission from ref. I.

The earlier results published on the strand-mimicking β-peptide oligomers revealed their tendency to self-assembling; consequently, we modeled the possibility of interstrand association for **4**. Parallel and antiparallel sheet-like dimers were built as starting geometries by using molecular mechanics, and these larger systems were optimized by using the force-field combined with an implicit solvent model. Both the parallel and the antiparallel arrangement resulted in stable dimers where the interstrand H-bonds are preferably positioned and there is no steric repulsion between the side-chains (Figure 11). These findings indicate the possibility of self-association into fibrillar morphology, because the dimer structures arrange the amide bond in the same direction, leading to a polar strand which has a strong tendency to associate. This structure was previously established only for hairpin-like oligomers.[51]

a

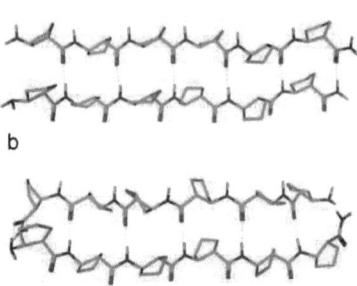
b

Figure 11. Geometry for the parallel (a) and antiparallel (b) dimers of **4**, showing the polar strand geometry, calculated by means of molecular mechanics (MMFF94x with implicit water) Reproduced with permission from ref. I.

The computational results support the view that new secondary structures can be created. As a consequence of these results, compounds **1-4** were synthesized. The amino acid coupling was carried out on a solid support, with Boc chemistry and DCC/HOBt as activating agents. In the case of **4**, difficulties appeared after the coupling of the fourth residue, which led to truncated products. The peptides were isolated by RP-HPLC. The final products were characterized by means of MS, various NMR methods, including COSY, TOCSY and ROESY, at 4 mM in CD_3OD, CD_3OH, DMSO-d_6 and water (90% H_2O + 10% D_2O) solutions, ECD in water and MeOH, VCD in DMSO and DCM.

The NMR signal dispersions were very good for **1** and **2**, and thus the resonance could be assigned along the backbone. However, **3** and **4** showed very low resonance dispersion; accordingly, no high-resolution structural information could be gained by NMR.

For measurement of the conformational stability of peptides **1** and **2**, NH/ND exchange was utilized in CD_3OD. The time dependence of the residual NH signal intensities of **2** demonstrated that the corresponding atoms are considerably shielded from the solvent due to H-bonding interactions (Figure 12). The exchange pattern observed is in good accordance with the theoretically predicted H10/12 helix. The

32

proton resonances of the N-terminal amine, the second amide and the C-terminal amide disappeared immediately after dissolution, while the other signals persisted for a longer period of time. For **1**, a similar exchange pattern was detected, but the exchange rates were significantly higher, suggesting a less-ordered structure.

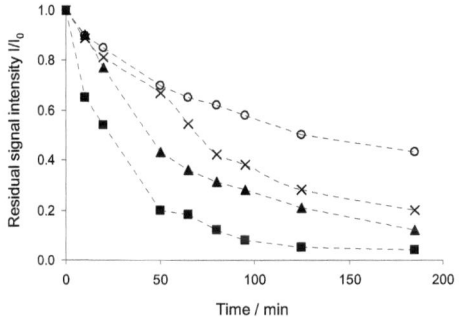

Figure 12. Time dependence of the NH/ND exchange for **2** in CD_3OD, ○ : NH_3; ▲ : NH_4; ■ : NH_5; x : NH_6 Reproduced with permission from ref. I.

Further evidence for the H10/12 helix was the pattern of the $C^\beta H$-NH vicinal couplings, which exhibited an alternating pattern (values given for DMSO): 9.8 Hz for NH_3-$C^\beta H_3$ and NH_5-$C^\beta H_5$, and 7.8 Hz for NH_2-$C^\beta H_2$, NH_4-$C^\beta H_4$ and NH_6-$C^\beta H_6$. These ^3J values are in good accordance with the *ab initio* H10/12 helical conformation.

To obtain direct evidence for the helical fold of **2**, ROESY experiments were run in CD_3OH, DMSO-d_6 and water (90% H_2O + 10% D_2O). In CD_3OH and DMSO-d_6, the characteristic $C^\beta H_2$-NH_4, NH_3-$C^\beta H_5$ and $C^\beta H_4$-NH_6 long-range NOE interactions could readily be observed, which reveal the predominant H10/12 helical conformation (Figure 13). For alternating peptide bond orientations and a helix conformation, the neighboring amide protons can be in proximity to each other. Accordingly, ROESY cross-peaks were detected for NH_3-NH_4 and NH_5-NH_6, which is further evidence in support of the H10/12 helix. Because of the nature of the H10/12 helix, the arrangement of the backbone hydrogens does not allow the observation of any further backbone NOE interactions.

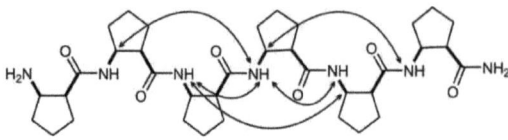

Figure 13. Long-range NOE interactions observed for **2.** Reproduced with
permission from ref. I.

In aqueous medium, these important cross-peaks and the alternating coupling
constant pattern did not appear. These observations point to the water-sensitivity of
the new H10/12 helix, but showed conformational stability even in the chaotropic
solvent DMSO. For **1**, only a weak $C^{\beta}H_2$-NH$_4$ NOE cross-peak was detected,
indicating the chain-length dependence of the helical structure.

For the complete solution-phase structure analysis, **2** was characterized by using
VCD in the chaotropic DMSO and the strongly H-bond-promoting and structure
supporting DCM. The recorded curves point to intense rotatory strengths in the amide
I and II regions of the vibrational spectrum in both solvents, and the theoretically
calculated VCD spectrum based on the H10/12 *ab initio* model is in very good
agreement with the experimental data with respect to both the frequencies and the
signs of the rotatory strengths, which not only proves the helical fold, but also yields
information on the absolute direction of the helicity. The VCD results in DMSO
corroborate that the H10/12 helix with cyclic side-chains remains stable even in the
H-bond-breaking solvent (Figure 14).

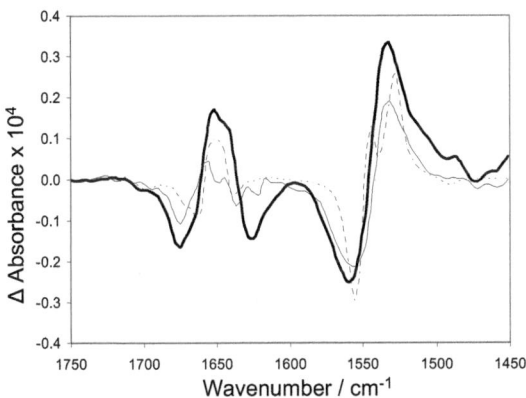

Figure 14. VCD curves for **2**, measured in DCM (thick continuous line), and in DMSO (thin continuous line), and the calculated results (dashed), given in scaled VCD units. Reproduced with permission from ref. I.

ECD measurements were carried out to gain further supporting evidence for the presence of the H10/12 helix. The ECD measurements in MeOH revealed a positive Cotton effect, with the positive band at ~ 210 nm and the negative band at ~ 190 nm for both **1** and **2** (Figure 15a), the intensity of the Cotton effect proving markedly lower for **1**. These findings support the view of the helically ordered structure in organic solvents in a chain-length-dependent manner. In water, the ECD curves (Figure 15b) displayed Cotton effects, but their intensities were considerably lower as compared with the ECD data in MeOH. The higher negative band at 190 nm for **1** indicates the presence of a more elongated conformation. Our H10/12 helix model with purely hydrophobic side-chains still exhibits some helical content in aqueous medium, but the structure is only partially ordered.

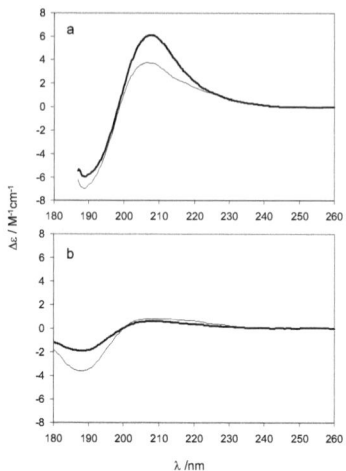

Figure 15. ECD results for **1** (thin curve) and **2** (thick curve) measured in MeOH (a) and in water (b), normalized to chromophore units. Reproduced with permission from ref. I.

The NMR measurements were not useful for **3** and **4** because of the signal dispersion problems. However, the absence of the well-resolved NMR signals does not automatically rule out the possibility of an ordered secondary structure. Moreover, the solubility of **4** was very low, preventing VCD measurements, where higher concentration is necessary. Consequently, the only accessible spectroscopic methodology with which to gain secondary structural information was ECD measurements. For **4**, the ECD curve furnished an intense negative Cotton effect (Figure 16), which suggests the presence of a periodic secondary structure. The negative band appeared with decreased relative intensity at ~ 205 nm, and the positive band was detected at ~ 190 nm. These features make the ECD curve similar to that observed for the homochiral *cis*-ACPC$_n$ strands (taking into account the absolute configuration).

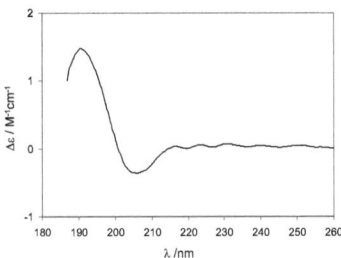

Figure 16. ECD results for **4**, measured in MeOH at a concentration of 1 mM, normalized to chromophore units. Reproduced with permission from ref. I.

The modeling results, the low solubility and the ECD measurements suggested the possibility of self-association and sheet formation in the case of **4**. Hence to test this, TEM measurements were performed in a search for any fibrillar morphology in solution. MeOH and aqueous solutions were made with concentrations of 4 mM. After incubation for 1 day at room temperature, no specific morphology was observed, but after 1 week in aqueous solution, the TEM images showed fibrils for **4**, with a length in the μm range and with an average width of 6 nm (Figure 17). The fibrils displayed uniform helicity, with a periodicity of 60 nm. In MeOH, no fibrillar morphology could be recognized, but some undefined aggregated objects were observed for **4**. Together with literature data on fibril-forming α-peptides, and previous results on the self-association of β-peptide strands into nano-ribbons, the TEM observations strongly suggest that the preferred secondary structure of **4** is the polar strand, which tends to self-assemble into sheets. Compound **3** did not show any self-assembly at all; this indicates that the self-association is chain-length-dependent.

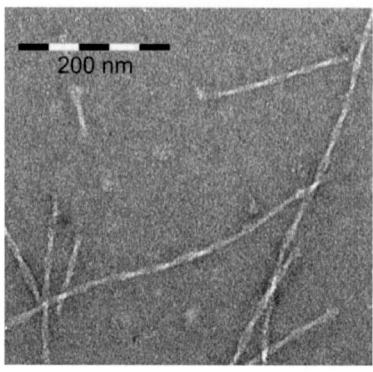

Figure 17. TEM image of **4** measured in water after 1 week of incubation. Reproduced with permission from ref. I.

4.2. Probing β-peptide H10/12 helix folding and self-assembly by side-chain shape

The H10/12 helix for β-peptide foldamers has been found to be intrinsically stable,[146] stabilized by the alternating heterochiral sequence of *cis*-ACPCs and β2/β3-peptides. Self-assembling phenomena have been reported for different β-peptide sequences leading to nanosized vesicles, helix bundles and lyotropic liquid crystals.[7] Controlled nanostructure formation with an H10/12 helix has never been observed, even with the strongly hydrophobic cyclopentane side-chain. Interestingly, this phenomenon has been reported only for H14 helices, especially with the *trans*-ACHC-containing sequences.[75,77,147]

To gain self-assembling H10/12 helices, our first initiative was to enhance the hydrophobic surface of the helix. Alternating heterochiral sequences were created with diverse 6-membered large and hydrophobic side-chains (Scheme 3.) with the aim of investigating the side-chain tolerance of the H10/12 helix and the self-assembling properties of different sequences.

Scheme 3. The structures with alternating backbone configurations made from *cis*-ACHC (**5, 6**), *cis*-ACHEC (**7, 8**) and *diexo*-ABHEC (**9, 10**). Reproduced with permission from ref. II.

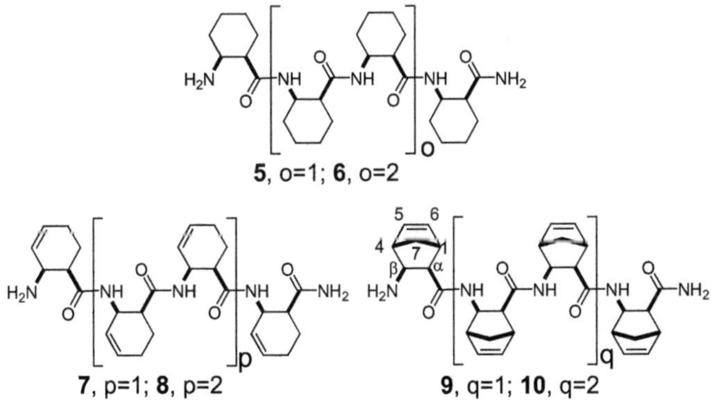

5, o=1; 6, o=2

7, p=1; 8, p=2   **9, q=1; 10, q=2**

β-Peptides were formed by the alternating coupling of *cis*-(1*R*,2*S*)-ACHC and *cis*-(1*S*,2*R*)-ACHC (**5, 6**), *cis*-(1*R*,2*S*)-2-aminocyclohex-3-enecarboxylic acid (*cis*-

(1*R*,2*S*)-ACHEC) and *cis*-(1*S*,2*R*)-2-aminocyclohex-3-enecarboxylic acid (*cis*-(1*S*,2*R*)-ACHEC), and *diexo*-(3*S*,4*R*)-3-aminobicyclo[2.2.1]hept-5-ene-2-carboxylic acid (*diexo*-(3*S*,4*R*)-ABHEC) and *diexo*-(3*R*,4*S*)-3-aminobicyclo[2.2.1]hept-5-ene-2-carboxylic acid (*diexo*-(3*R*,4*S*)-ABHEC) through fine-tuning of the β-peptide side-chain shape.

The foldamers were synthesized by solid-phase peptide synthesis with Fmoc technology, leading to unprotected sequences. The crude peptides were purified by RP-HPLC. The saturated oligomers **5** and **6** were prepared in two ways: (*i*) the catalytic reduction of **7** and **8** in an H-Cube® apparatus, and (*ii*) direct coupling of the Fmoc-*cis*-ACHC monomers. The pure peptides were characterized by means of RP-HPLC, MS and various NMR measurements (COSY, TOCSY and ROESY), with different solvents: 4 mM solutions in CD_3OH, DMSO-d_6, and water (H_2O/D_2O 90:10). The NMR signal dispersions were good for most of the compounds in these solvents; no signal broadening was observed and signal assignment could be performed. Interestingly, the only exception was **6**, where significant signal broadening was detected in all solvents. However, cooling of the sample to 245 K in $CDCl_3$ resulted in well-resolved signals, where two sets of signals were frozen out. The signals could be assigned only for the major conformer. This suggested that **6** has two distinct conformations that are in rapid chemical exchange.

For assessment of the conformational stabilities, NH/ND exchange was studied in CD_3OD. For **8** and **10**, the proton resonances relating to the terminal nitrogen, the amide NH_2 and the C-terminal amide disappeared directly after dissolution, while other signals persisted even 1 h after dissolution (Figure 18). For **6**, signal assignment was not possible, and consequently the average signal intensity was investigated as a function of time. This exhibited a similar pattern as for **8** and **10**. These results indicated the existence of a folded structure stabilized by H-bonds because the amide protons are considerably shielded from the solvent. The shorter sequences **5**, **7** and **9** exhibited significantly higher exchange rates; all the signals were practically lost within 40 min. This suggests the chain-length-dependent property of the oligomers, because increasing chain length leads to decreased exchange rates, which is caused by increased structural stability.

To achieve further low-resolution structural data, ECD spectra were recorded at room temperature in MeOH and water at 1 mM (Figure 19). From a comparison with the ECD spectrum of the H10/12 helix of **2**, the ECD spectra of **6** and **8** in MeOH indicated an H10/12 helix with the opposite helicity. The signal intensities for **6** and **8** were lower, indicating a destabilizing effect of the 6-membered side-chains. This structure was in part also retained for **8** in water. The ECD curve of **6** showed a significant drift in water, possibly due to the light scattering of the large aggregates (see DLS and TEM results), and this data set could therefore not be interpreted.

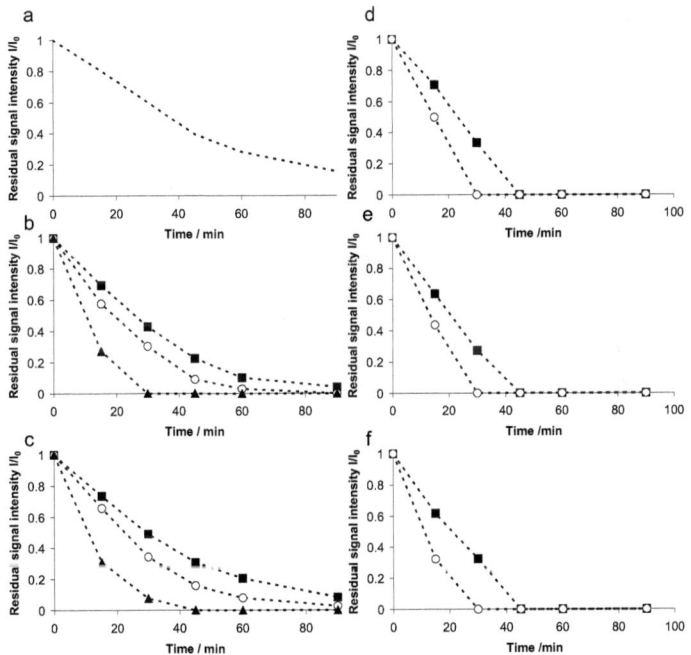

Figure 18. Time dependence of the NH/ND exchange measured in CD$_3$OD for the hexameric structures **6** (a), **8** (b) and **10** (c); ■: NH$_3$; ○: NH$_4$; ▲: NH$_5$; and for the tetrameric sequences **5** (d), **7** (e) and **9** (f); ■: NH$_2$; ○: NH$_3$. Reproduced with permission from ref. II.

The ECD curve measured for **10** in MeOH showed a negative Cotton effect with a minimum at ~ 215 nm and a maximum at ~ 200 nm. This clear difference from the typical ECD curve of the H10/12 helix and also from the known ECD patterns of β-peptide strand-like structures supports[53] the view that **10** does not fold into the H10/12 helix in the solution phase, but the overall structure exhibits helicity. The ECD curve of **10** measured in water demonstrated a clear difference from that in MeOH with a positive Cotton effect with a maximum at ~ 210 nm and a minimum at ~ 190 nm. These significant changes indicate a solvent-driven structural change.

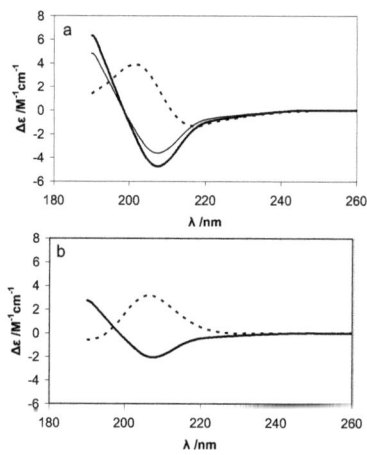

Figure 19. The ECD curves of 1 mM **6** (thin), **8** (thick) and **10** (dashed) solutions in methanol (a) and of 1 mM **8** (thick) and **10** (dashed) solutions in water (b). Reproduced with permission from ref. II.

To acquire high-resolution structural data, NMR measurements were performed. Compound **6** exhibited good resonance dispersion in CDCl$_3$ at 245 K, where two chemically exchanging conformers were observed. Signal assignment was performed for the major conformer, and the NOESY spectrum at 245 K revealed a long-range NOE pattern typical of the left-handed H10/12 helix: NH$_1$-C$^\beta$H$_3$, C$^\beta$H$_2$-NH$_4$, NH$_3$-C$^\beta$H$_5$ and C$^\beta$H$_4$-NH$_6$ (Figure 20a). Structure refinement was performed by using distance restraints derived from the NOESY cross-peaks, and the low-energy conformational cluster corresponds to the left-handed H10/12 helical structure. This conformation was subjected to *ab initio* quantum chemical optimization. At both the RHF/3-21G and B3LYP/6-311G** levels of theory, the left-handed H10/12 proved stable (Figure 21a). Resonance assignment could not be performed for the minor conformer because of its low relative abundance (~ 6%). The low overall NH/ND exchange rate for **6** indicates that the conformational equilibrium can not involve an unfolded minor conformer or a significant fraction of unfolded intermediate states. Molecular modeling at the B3LYP/6-311G** level suggested that the H-bond-stabilized conformation closest in energy is the left-handed H18/20 ($\Delta E = 4$ kcal mol^{-1}). The left-handed H18/20 helix is proposed, which can occur through a rapid shift between the H-bonds, without overall disassembly of the helical structure. For longer sequences, this helix could be the predominant conformation.

For **8**, the long-range NOE pattern observed in the ROESY spectra in CD$_3$OH and DMSO-d_6 revealed the presence of an H10/12 helix, and the structure refinement resulted in the left-handed H10/12 as well (Figure 21b). Further optimization at the RHF/3-21G and B3LYP/6-311G** levels of theory revealed the stability of the H10/12 helix.

For **10**, with backbone protecting groups, a bent structure has previously been described.[148] The ROESY spectra of **10** showed similar cross-peak patterns in the different solvents. A series of NOE interactions were found for the pairs NH$_i$-C^{4}H$_i$ and

NH$_i$-C^{1}H$_{i-1}$, but $i - (i+2)$ interactions could not be detected (Figure 20c). This NOE pattern excludes formation of the alternating H10/12 helix. NMR-derived distance restraints-based force field calculations indicated the formation of a circular and not a bend-like fold stabilized by 6-membered H-bonded pseudorings, where all the large side-chains are on the same side of the molecule (Figure 21c). Furthermore, the absence of C^{7}H$_i$-C^{1}H$_{i-1}$ and C^{7}H$_i$-C^{4}H$_{i+1}$ NOE interactions in the ROESY spectra, which should appear for an elongated strand, point to the circular fold. The *ab initio* quantum chemical calculations at both the RHF/3-21G and B3LYP/6-311G** levels of theory supported this conformation (Figure 21c). The analysis of the folded structures of **2**, **6**, **8** and **10** revealed that, for a H10/12 helical fold, a θ of ~ 30° is necessary. For the bridged bicyclic side-chain θ is constricted to 0°, excluding the formation of a helical structure and leading to a circular fold.

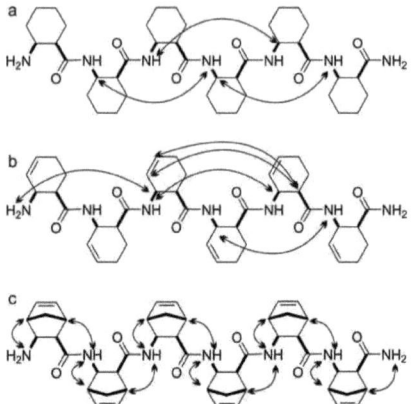

Figure 20. Long-range NOE interactions for **6** (a) in CDCl$_3$ and for **8** (b) and for **10** (c) in DMSO and CD$_3$OH. Reproduced with permission from ref. II.

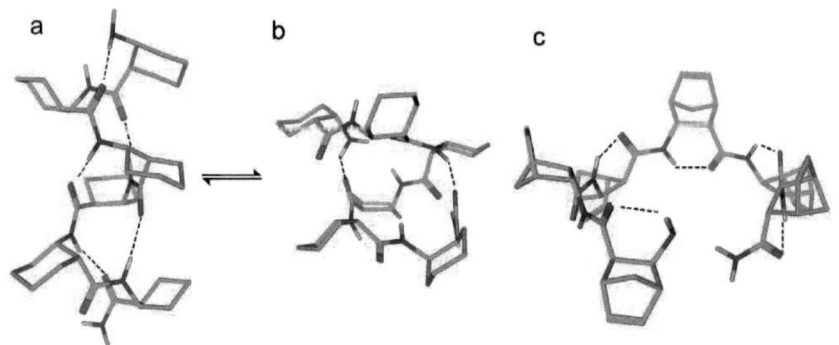

Figure 21. The H10/12 helix geometry for **6** (a), the H18/20 helix geometry for **6** (b) and the circular fold for **10** (c), obtained through NMR restrained structure refinement and by final optimization at the B3LYP/6-311G** level of theory. Reproduced with permission from ref. II.

For analysis of the self-association, TEM and DLS were used. For a 4 mM solution of **6** in water, vesicles with an average diameter of 100 nm were observed in the TEM images immediately after dissolution, sonication and filtration (Figure 22).

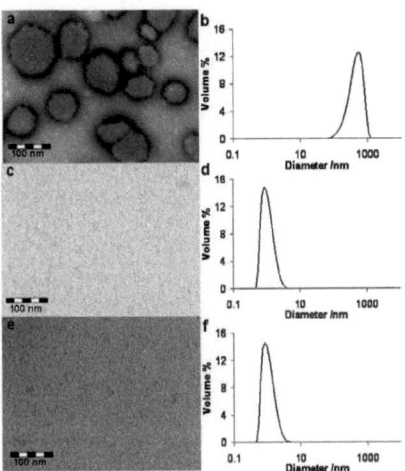

Figure 22. TEM images of **6** (a), **8** (c) and **10** (e) and DLS curves for **6** (b), **8** (d) and **10** (f) as 4 mM solutions in water. Reproduced with permission from ref. II.

The vesicles showed cluster formation by self-association. DLS measurements supported this result and indicated particles with diameters in the range 100-1000 nm. For **5** and **7-10**, no self-association was observed under the same conditions. A minor alteration in the side-chain shape resulted in considerably different behavior in the association properties of the β-peptide foldamers **2** and **5-10**. Solvent-driven self-assembly of the H10/12 helix was not observed with cyclopentane (**2**) and cyclohexene (**8**) side-chains; only the cyclohexane side-chain (**6**) led to vesicle formation in water. This is an indication that the moderately distinct hydrophobic nature of the cyclohexane side-chain supports self-assembly. On the other hand, the bulky and hydrophobic *cis*-ABHEC residue in itself is not enough to promote aggregation, indicating that the fold into the compact helical structure is also needed. Consequently, collaboration of the stable compact helical conformation and the sufficiently hydrophobic side-chain is necessary for the formation of self-assembly. For an assessment of the hydrophilic/hydrophobic character of the studied foldamers in their folded state (lowest energy *ab initio* geometries), hydrophobic and hydrophilic surface areas were calculated (Table 1).

Table 1. Surface areas calculated for β-peptide oligomers with an alternating backbone configuration and the indicated side-chains. Reproduced with permission from ref. II.

Cpd.	Side-chain	Hydrophobic surface (Å²)	Hydrophilic surface (Å²)	Ratio
6		802.7	142.3	5.6
8		690.5	250.2	2.8
10		759.9	283.7	2.7
2		726.1	147.5	4.9

The calculated values indicate that the self-assembling foldamer **6** exhibits the largest hydrophilic/hydrophobic surface ratio in its folded state.

4.3. Stereochemical patterning approach for *de novo* foldamer construction

Efficient design of the secondary structures of the peptidic foldamers is a great challenge. The process mainly relies on human intuition or trial and error experiments and handles merely a very limited subset of the numerous possible parameters of foldamer design: the pattern of α- and β-residues, the side-chain topology/chemistry, the backbone configurations, the side-chain interactions, etc. A highly important aspect of the rational control over the secondary structure of the peptidic foldamers is the stereochemistry of the amino acid building blocks. The effects of the configurations at the backbone carbon atoms on the overall geometry are conveyed through the local torsional preferences and the side-chain – backbone interactions. The literature findings discussed in section 2.2.3. reveal that the stereochemistry of the amino acid monomers effectively determines the prevailing secondary structure. The previous two sections showed that the extension of the variation of the backbone stereochemistry to the dimer level results in new helical and strand-like secondary structures. Taking these results into account, the generalization of the configuration patterning along the backbone and the simultaneous variation of the residue type (α-/β-residues) can lead to novel periodic secondary structures. Furthermore, the construction of a rule which facilitates an understanding of the effect of the stereochemistry on the prevailing secondary structure can yield a simple, intuitive foldamer design tool and furnishes highly effective *de novo* construction methodology for the generation of peptidic foldamers.

The effect of the backbone configuration on the geometry of the foldamer is transmitted by the dihedrals and the side-chain – backbone interactions. In order to test the effects of the stereochemical patterns governing the folding properties, a set of H-bond-stabilized periodic secondary structures were analyzed and the backbone dihedrals were investigated (Table 2). Mainly the φ and ψ dihedrals were taken into account, but the β-residues contain the θ dihedral, which is dependent on the φ and ψ dihedrals and its gauche conformation is intrinsically stable.[149,150]

Table 2. Representative backbone torsions and the corresponding absolute configurations of the known secondary structures built up from α- and β-amino acids. Reproduced with permission from ref. III.

Residues	Secondary structure	$[\varphi,\psi]$ repeating units in degrees[b]	Absolute configurations[c]	Periodicity[d]
α	α-helix[151]	$[-65,-39]^{\alpha}$	$[S]^{\alpha}$	4
α	3_{10}-helix[151]	$[-50,-25]^{\alpha}$	$[S]^{\alpha}$	3
α	π-helix[151]	$[-45,-80]^{\alpha}$	$[S]^{\alpha}$	5
α	β-helix[152]	$[-153,144]^{\alpha}[125,-124]^{\alpha}$	$[S]^{\alpha}[R]^{\alpha}$	6
β	H14 helix[21,23]	$([138,135]\beta)_n$	$[RR]^{\beta}$ $[RX]^{\beta}$	3
β	H12 helix[36]	$[-92,-105]^{\beta}$	$[SS]^{\beta}$	2
β	H10 helix[153]	$[89,83]^{\beta}$	$[RR]^{\beta}$	2
β	H8 helix[154]	$[-90,-53]^{\beta}$	$[SS]^{\beta}$	1
β	H10/12 helix[24,25]	$[91,-99]^{\beta}[-108,91]^{\beta}$ $[-90,100]^{\beta}[100,-90]^{\beta}$	$[RS]^{\beta}[SR]^{\beta}$ $[SX]^{\beta}[XS]^{\beta}$	2
αβ	H14/15 helix[55]	$[77,31]^{\alpha}[140,108]^{\beta}$	$[R]^{\alpha}[RR]^{\beta}$ $[R]^{\alpha}[RX]^{\beta}$	3
αβ	H11 helix[55]	$[58,35]^{\alpha}[94,83]^{\beta}$	$[R]^{\alpha}[RR]^{\beta}$ $[R]^{\alpha}[RX]^{\beta}$	2
αβ	H9/11 helix[56,57]	$[-70,125]^{\alpha}[100,-76]^{\beta}$ $[68,-128]^{\alpha}[-102,72]^{\beta}$	$[S]^{\alpha}[RS]^{\beta}$ $[R]^{\alpha}[SX]^{\beta}$	2
$\alpha\beta_2$	H11/11/12 helix[59]	$[-57,-35]^{\alpha}[-101,-91]^{\beta}_2$	$[X]^{\alpha}[SS]^{\beta}_2$	2
$\alpha_2\beta$	H10/11/11 helix[59]	$[-54,-31]^{\alpha}_2[-99,-85]^{\beta}$	$[X]^{\alpha}_2[SS]^{\beta}$	2
α	β-strand	$[-120,113]^{\alpha}$	$[S]^{\alpha}$	a
α	α-strand (DL-stabilized)[66]	$[-50,-50]^{\alpha}[50,50]^{\alpha}$	$[S]^{\alpha}[R]^{\alpha}$	a
β	extended polar	$[144,-154]^{\beta}$	$[RS]^{\beta}$	a

	strand[51]			
β	Z6 non-polar strand[53]	[-160,75][β]	[SR][β]	a
β	alternating polar strand	[147,170][β][-126,-163][β]	[SS][β][RR][β]	a

[a]Not applicable
[b]Superscripts indicate the corresponding residue type.
[c]X stands for the unsubstituted (CH_2) or the symmetrically disubstituted ($-C(CH_3)_2-$, Aib) centers
[d]Number of peptide bonds (n) between ($i - i+n$) H-bonds

The analysis revealed that the peptide bond-flanking dihedrals ($\psi_{(n-1)}$][φ_n, where][designates the amide group) have the same sign for helical structures, while they have opposite signs for strand-like structures. The results are summarized in Figure 23.

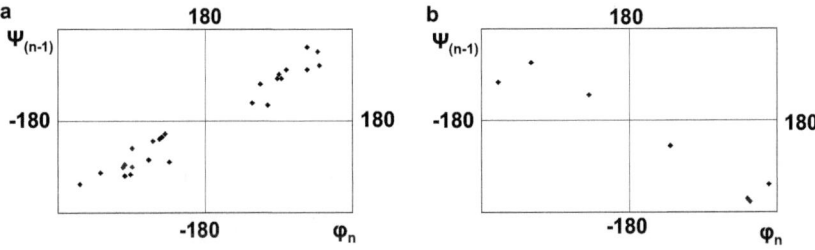

Figure 23. Plots of different helices (a) and strands (b) where the amide bond-flanking φ_n and $\psi_{(n-1)}$ dihedrals are shown. Reproduced with permission from ref. III.

Periodic secondary structures, and especially helices, have a fixed general periodicity along a sequence and the structure requires H-bonding network for stabilization with accurately positioned amide bonds at periodic distances. Consequently, the order of the φ and ψ dihedrals must also exhibit a periodic pattern. If a helix is stabilized by $i - (i+n)$ peptide bond contacts, the φ and ψ pattern starting from position i should be identical to that starting at $i+n$. The homochiral helices with uniformly oriented amides and the alternating heterochiral structures with alternating amide orientations are special cases of the class of periodic φ and ψ sign sequences.

Consequently, the crucial point is the combined orientation of the peptide bond caused by the flanking stereocenters.

The sign of φ and ψ can be efficiently tuned by the backbone stereochemistry.[155,156] The stereochemistry can easily be correlated with the CIP (Cahn-Ingold-Prelog) convention by applying side-chain replacement (hypothetically changing all side-chains with a methyl group), retaining the same designation for the same stereochemistry (Table 3). The backbone configurations strongly and independently restrain the peptide bond orientation relative to the backbone and side-chains on both the N and C sides, and the stereochemistry in the β-peptidic backbone is conveyed effectively to the folding propensies of the β-peptide foldamers (see Figure 9). For an α-residue, there is only a single stereocenter in the backbone, and the C=O can point either to the identical side with H_α in the α-, 3_{10}- and π-helices[151] or to the opposite side for the β-strand and β-helix.[152] Thus, the structuring effect on the peptide bond on the carbonyl side is significantly weaker for the α-residues. Sequences containing unsubstituted or disubstituted backbone atoms tend to adopt a helical conformation.[157]

Table 3. Signs of dihedral angles ($[\varphi_n$ and $\psi_{(n-1)}]$) induced by the stereochemistry of the backbone. Reproduced with permission from ref. III.

Residue type	Backbone configuration	Structuring effect
α	(S)-C_α	[−
α	(R)-C_α	[+
β	(S)-C_α	[−
β	(R)-C_α	[+
β	(S)-C_β	−]
β	(R)-C_β	+]

The above relationships were deduced from a descriptive system, and the interesting question arises as to whether these rules have any predictive ability. To test this rule in a *de novo* helix design, $[\varphi_n$ and $\psi_{(n-1)}]$ sequences and the corresponding stereochemical patterns were created. The stereochemical patterning was extended to the whole sequence and the designed streams were translated into real β- and α/β-peptide foldamers by using ACPC diastereomers, homologated β_3-amino acids and α-amino acid enantiomers with appropriate absolute configurations (Figure 24). For conformational constraints, the cyclopentyl residues were selected because they have been reported as only possible cyclic β-amino acid combination with α-amino acids.[13] The side-chain chemistry was chosen so as to permit advantageous interactions between residues in potential juxtaposition, giving extra stabilization for the prevailing helical structures, and to introduce biomimetic proteinogenic side-chains leading to increased solubility and opening the way for future biological applications. Sequences **11-13** were created so as to have the same $[\varphi_n$ and $\psi_{(n-1)}]$ signs, while **14-16** are the control sequences and were constructed so to have $[\varphi_n$ and $\psi_{(n-1)}]$ with opposite signs in certain positions to disrupt the periodic pattern in the central region. Consequently, compounds **11-13** are anticipated to form a helical structure and **14-16** are predicted to have a disrupted non-helical structure.

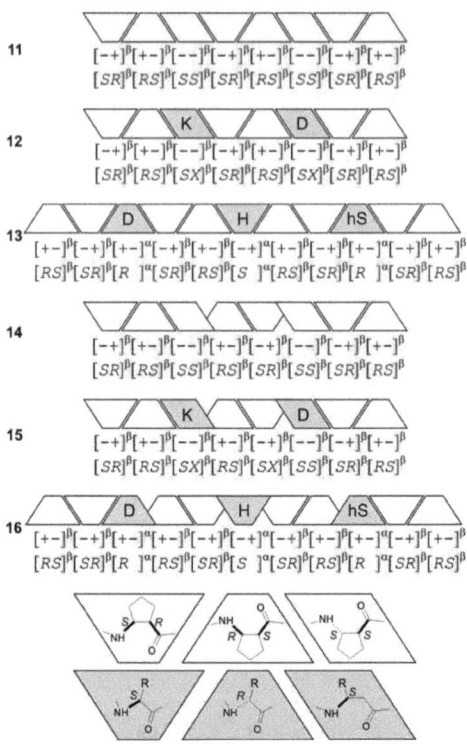

Figure 24. The *de novo* designed β- and α/β-peptide sequences based on the [φ,ψ] sign and stereochemical patterns (**11-13**) and their quasi-randomized controls (**14-16**). The side-chains are indicated by the single letter codes (hS: homo-serine). The configurations are indicated in the CIP system with R = methyl. Reproduced with permission from ref. III.

First, a conformational search was carried out by using a hybrid MC/MD calculation with the MMFF94x force field without any distance restraint. The conformational pool was clustered and the resulting lowest-energy conformational ensembles for **11-13** indicated helical conformations. For **14-16**, turn-like segments were detected in the low-energy conformations, which lacked any long-range order. The lowest-energy conformers from simulations of **11-13** were further optimized at the *ab initio* quantum chemical level. The computation was performed first at the RHF/3-21G level of theory, with further structure refinement at the B3LYP/6-311G** level of theory. The structure optimizations converged properly, and novel foldameric helix types were found. For these structures, a further single point energy calculation was performed with the PCM. For **11** and **12**, the left-handed (*M*) helix is stabilized by concatenated 14- and 16-membered H-bonded pseudorings, whereas **12** exhibits a potential salt-bridge interaction between the side-chains in juxtaposition (H14/16, Figure 25a). For **13**, the left-handed (*M*) helix is stabilized by consecutively concatenated 9-, 10-, 11- and 12-membered pseudorings in a repeating unit (H9-12; Figure 25b). This network is augmented by the H-bonds between side-chains in juxtaposition, which introduces a slight curvature into the helix, but does not disrupt the backbone H-bonds. The peptide bond orientations display a regular alternating pattern, which indicates that stereochemical pattern compatibility is a crucial factor in the self-recognition of the helical turns. The amide orientations display the designed pattern in all the structures.

a   b

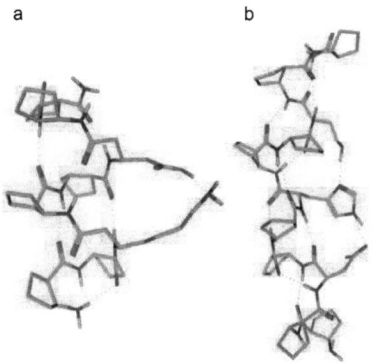

Figure 25. (a) The H14/16 helix obtained by molecular modeling for **12**; (b) the H9-12 helical conformation gained for **13**. Reproduced with permission from ref. III.

In view of the possible solubility problems and with the aim of the incorporation of proteinogenic side-chains, compounds **12**, **13**, **15** and **16** were synthesized on a solid support. The purification was performed by RP-HPLC and the products were characterized by analytical HPLC, MS and various NMR measurements at 8 mM in CD$_3$OH, DMSO-d_6 and water (H$_2$O:D$_2$O 90:10) solutions. The NMR signal dispersions were good for **12** and **13**; no resonance broadening was observed. The signal assignment was carried out at 277 K, where the best-resolved spectra were obtained. As expected for the disrupted structures **15** and **16**, the signal dispersions were lower even at reduced temperature, indicating the absence of the periodic secondary structures.

For assessment of the conformational stabilities, NH/ND exchange was studied. Interestingly, the signals of the amide protons disappeared immediately after dissolution. This phenomenon can be explained by the presence of acyclic charged side-chains. This leads to faster exchange. This effect can be excluded by measuring the backbone amide H_N chemical shift temperature gradients ($\Delta\delta \, \Delta T^{-1}$ values) in DMSO-d_6. With this measurement, the presence of folded structures stabilized by H-bonds was tested.[158] For **12** and **13**, the temperature gradients were generally less negative than -5 ppb K^{-1} and decreasing trends were observed on going from the terminal residues to the central part of the sequences, with the lowest negative value at ~ -4 ppb K^{-1} (Figure 26). These results demonstrate considerable shielding from the solvent, indicating the presence of an H-bond-stabilized secondary structure. Sequence **15** displays temperature coefficients < -5 ppb K^{-1} in the disrupted central region. For **16**, the informative resonance assignment was also partially possible; nevertheless, the disrupted central region gives coefficients < -5 ppb K^{-1}. Interestingly, strong H-bonds are observed in the NH3 and NH10 positions, indicating some partially ordered structure. The results reveal that **15** and **16** exhibit values relating to random or partially organized structures with H_N protons mostly exposed to the solvent.

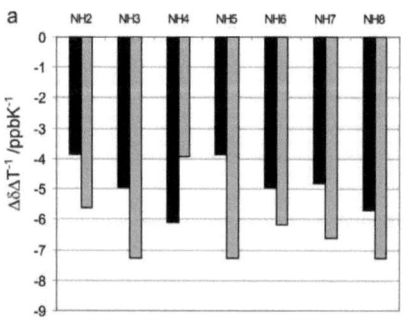 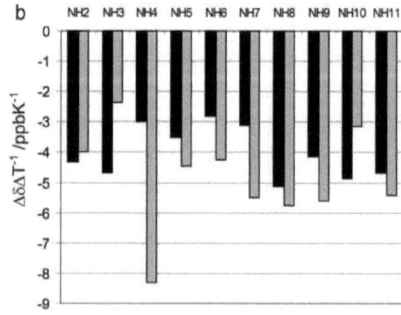

Figure 26. Temperature gradients of the chemical shifts of the amide protons measured in DMSO-d_6. Panel (a) for **12** (black) and **15** (gray), and panel (b) for **13** (black) and **16** (gray). Reproduced with permission from ref. III.

For further structural fingerprints and to monitor the changes in the overall fold of the synthesized models, ECD measurements were carried out (Figure 27). The spectra were run at room temperature and, for consistence with the NMR analysis, at 277 K too. The two sets of spectral data did not exhibit a significant difference upon decrease of the temperature, apart from the slightly increased intensities for **12** and **13**.

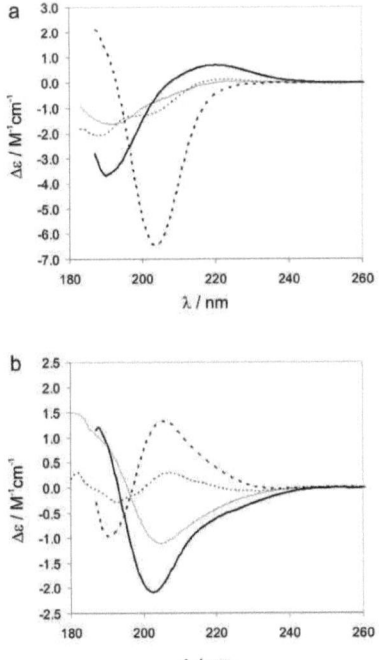

Figure 27. ECD curves measured at room temperature in MeOH (thick) and water (thin). Panel (a) for **12** (continuous) and **15** (dashed). Panel (b) for **13** (continuous) and **16** (dashed). Data were normalized to unit chromophore. Reproduced with permission from ref. III.

A positive Cotton effect was detected for **12**, with a positive lobe at around 220 nm and a negative band at 190 nm. The disrupted structure **15** exhibits a largely

55

asymmetrical low-wavelength Cotton effect (negative lobe at 205 nm) with opposite sign as compared with the parent structure **12**. The asymmetric behavior suggests elongated geometry. The large change in the ECD fingerprint indicates that the exchange of the two central β-residues practically destroys the H14/16 helix, which is in line with the modeling results. For **13**, a negative Cotton effect was observed, with negative and positive lobes at 205 nm and 190 nm, respectively. Literature data on the ECD spectra of the β-peptidic helices[1,6,8] suggest that the shift of the higher-wavelength band is related to the increasing pitch height, which is indeed apparent from our data for **12** (H14/16) and **13** (H9-12). For **16**, the disruption caused an inverted ECD curve relative to the parent sequence **13**. The inverted, but still existing symmetric Cotton effect can be attributed to the local organization of the turn-like motifs observed at the molecular modeling stage. The ECD curves recorded in water furnished further convincing evidence of the stability of the designed helices. Pairwise comparison of the spectral features demonstrates that the designed helix **12** partially retains its ECD features even in the H-bond-destroying aqueous environment, indicating the same H9-12 helix. Compound **13** and the disrupted structures (**15** and **16**) display either a great intensity loss or a complete difference meaning extensive structural loss or change.

To gain high-resolution structural data, NMR (COSY, TOCSY and ROESY) measurements were performed. The resolved long-range NOE interactions in the ROESY experiments for **12** and **13** are indicated in Figure 28.

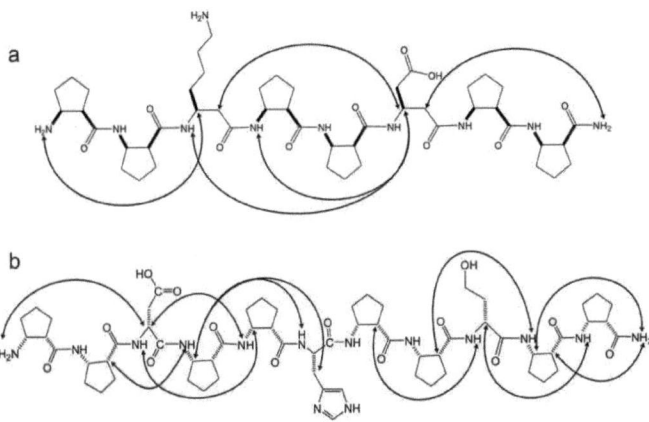

Figure 28. Resolved long-range NOE interactions for **12** (a) and **13** (b). Reproduced with permission from ref. III.

For **12**, the $i – (i+2)$ and $i – (i+3)$ interactions are in very good agreement with the predicted self-organization pattern. The ROESY cross-peaks related to the N- and C-terminal parts of the sequence exhibited measurable, but weak signals, which suggests terminal fraying. Nevertheless, the redundant NOE interactions in the central part prove H14/16 as the major conformer. It is important that strong NOEs were observed for $C^{\alpha}H_3$-NH_4 and $C^{\alpha}H_6$-NH_7, where $C^{\alpha}H_n$ indicates the axially oriented protons exhibiting long-range interactions along the helix. This finding strongly supports the view that the local geometry around the unsubstituted backbone atoms accommodated the H14/16 helix. The NH-$C^{\beta}H$ and NH-$C^{\alpha}H$ vicinal couplings were uniformly > 8.5 Hz, indicating the antiperiplanar orientation of the corresponding protons, which is necessary for the helical conformation. Further molecular dynamics studies with the NMR restraints revealed that exclusively the H14/16 fold can satisfy the observed NOE signals, allowing the conformational equilibrium of the fraying terminals.

Sequence **13** exhibited a signal rich ROESY spectrum with a number of $i - i+2$ interactions at exactly the same positions as predicted for the H9-12 conformation. In this case, signs of terminal fraying could not be detected in the spectrum. Fortunately, a $C^{\beta}H_4 - C^{\beta}H_6$ backbone – side-chain NOE interaction could be resolved. The interaction occurs only for the one geminal proton on the His_6 side-chain, which is antiperiplanar to the $C^{\alpha}H_6$ ($^3J = 9.53$ Hz), thereby corroborating that the side-chain orientation is the same as predicted. The $NH-C^{\beta}H$ vicinal couplings measured on the β-residues were > 9.0 Hz, while $^3J(NH-C^{\alpha}H)$ for α-residues was ~ 7.5 Hz. These values are in line with the H9-12 helix, because it requires that the α-residues attain a smaller $NH-C^{\alpha}H$ torsion than β-residues at the $NH-C^{\beta}H$ dihedral. For **15** and **16**, mainly sequential and intra residue NOE peaks were found.

Further structure refinement was performed by using hybrid MC/MD simulation with NMR-derived distance restraints. The results are presented in Figure 29.

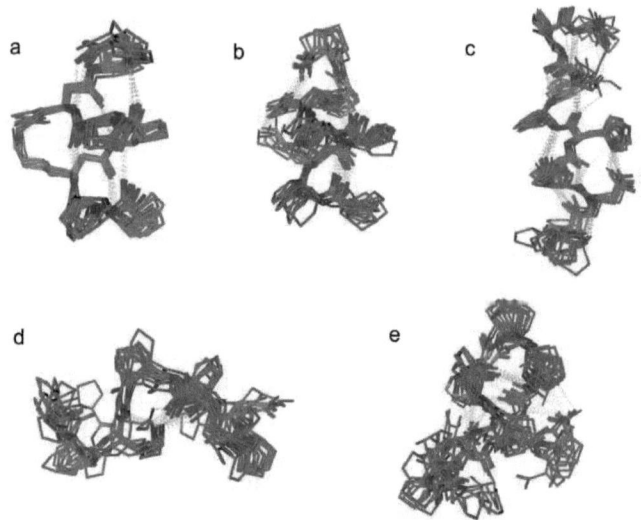

Figure 29. The 10 lowest energy structures obtained from the NMR structure refinement for **12:** a (cluster 1) and b (cluster 2), **13**: c, **15**: d and **16**: e. Reproduced with permission from ref. III.

The cluster analysis for **12** revealed two distinct conformational ensembles, the difference between them being the fraying of the C-terminal, which is in good agreement with the NMR results. For **13,** only one conformational family was gained, which corresponds to the H9-12 helix. For the structure refinement of **15** and **16**, no long-range NOEs were used and the calculation resulted in loop-like geometry for both compounds.

5. Summary

On the basis of earlier results, we performed the molecular modeling of various peptides with heterochiral backbones and different side-chains. After careful selection of the resulting conformational ensemble we synthesized the compounds with ordered structures. Peptides **1-10**, **12**, **13**, **15** and **16** were prepared by Boc or Fmoc techniques. The purification was performed by RP-HPLC. The peptides were characterized by analytical HPLC and MS. The structures of the synthesized peptides were determined in different solvents through the use of various NMR techniques, ECD and VCD.

The combination of the alternating design principle with the heterochiral sequences led to foldamers with novel 3D structures and the accessible conformational space of foldameric secondary and tertiary structures was enlarged.

The alternating heterochiral *cis*-ACPC hexameric **2** formed the H10/12 helix. For the alternating heterochiral *cis*-ACPC oligomers, no self-assembly was observed.

For the alternating heterochiral *trans*-ACPC sequences, molecular modeling and ECD results indicated a polar-strand conformation, which was supported by the fact that self-assembly into nanostructured fibrils was observed in water for the hexameric structure **4**. The self-assembly was chain-length-dependent, since tetramer **3** did not exhibit self-association.

The β-peptidic alternating H10/12 helix tolerates the 6-membered side-chain topology. Both the *cis*-ACHC hexamer **6** and the *cis*-ACHEC hexamer **8** afforded the H10/12 helix. The ECD results indicated that the 6-membered ring topology slightly destabilizes the H10/12 helix as compared with the alternating *cis*-ACPC oligomers.

The *cis*-ACHC-containing hexamer **6** with alternating backbone configuration exhibited conformational polymorphism with two folded conformational states that underwent chemical exchange. This behavior was not observed for the double bond-containing *cis*-ACHEC hexamer **8**. An apparently subtle change in the hybridization of a carbon atom pair in the side-chain can tune the conformational preferences of β-peptide oligomers.

The bicyclic *diexo*-ABHEC prevented the formation of a small-diameter helix; the experimental results pointed to a circle-like fold for the hexamer **10**, which was sufficiently stable in MeOH to maintain considerably shielded H-bonds.

Our results reveal that, for the β-peptide H10/12 helix studied, minor changes in the side-chain topology, size and shape resulted in large effects on the self-assembly process. The H10/12 helix is capable of self-association in a polar medium if the helix is built up from *cis*-ACHC residues and the chain length is as long as the hexamer. We showed that the hydrophobically driven helix association is a result of the combination of the hydrophobic nature of the side-chains and the presence of a stable secondary structure.

We analyzed various secondary structures of different foldameric and α-peptidic structures as a function of the back-bone stereochemical pattern. The results showed that the homochiral and the alternating heterochiral systems do not cover all the possibilities to create periodic secondary structures.

The absolute configurations can be regarded as the basic instruction set in the assembly language of the peptidic foldamer sequences, and an SPA was established.

We tested the SPA experimentally in terms of sequences with a novel backbone stereochemical pattern, giving an H14/16 helix for compound **12** and an H9-12 helix for oligomer **13**. These are *de novo* designed helices. Further evidence in connection with the relevance of the SPA is the two distorted structures (**15** and **16**) with the swapped backbone pattern, because they do not form helical structures. Their secondary structure is loop-like.

Acknowledgments

I am grateful to my supervisor, Prof. Ferenc Fülöp, Head of the Institute of Pharmaceutical Chemistry, University of Szeged, for his guidance of my work, his inspiring ideas, his useful advice and his constructive criticism.

My special thanks are also due to my co-supervisor, Dr. Tamás Martinek, for his continuous support, his unrestricted supervision of my work and his constructive criticism as well.

I would like to thank Dr. Livia Fülöp for helpful discussions and the DLS and TEM measurements.

I express many thanks to Dr Anasztázia Hetényi for her criticism in general and for reading my thesis carefully.

I am also grateful to all my colleagues and friends, especially Edit Wéber, Dr. Árpád Balázs, Dr. Gábor Tasnádi and Sándor Ötvös, for their practical advice and inspiring discussions.

My special thanks go to my family, for their inexhaustible support.

References

1. Cheng, R. P.; Gellman, S. H.; DeGrado, W. F. *Chem. Rev.* **2001**, *101*, 3219.

2. Gellman, S. H. *Acc. Chem. Res.* **1998**, *31*, 173.

3. Martinek, T. A.; Fülöp, F. *Eur. J. Biochem.* **2003**, *270*, 3657.

4. Fülöp, F.; Martinek, T. A.; Tóth, G. K. *Chem. Soc. Rev.* **2006**, *35*, 323.

5. Seebach, D.; Matthews, J. L. *Chem. Commun.* **1997**, 2015.

6. Seebach, D.; Beck, A. K.; Bierbaum, D. J. *Chem. Biodev.* **2004**, *1*, 1111.

7. Goodman, C. M.; Choi, S.; Shandler, S.; DeGrado, W. F. *Nat. Chem. Biol.* **2007**, *3*, 252.

8. Hecht, S.; Huc, I. (Eds) Foldamers: Structure, properties and applications: foldamers based on remote intrastrand interactions Wiley-VCH, Weinheim, **2007**.

9. Seebach, D.; Gardiner, J. *Acc. Chem. Res.* **2009**, *41*, 1366.

10. Sharma, G. V. M.; Reddy, K. R.; Krishna, P. R.; Sankar, A. R.; Narsimulu, K.; Kumar; S. K.; Jayaprakash, P.; Jagannadh, B.; Kunwar, A. C. *J. Am. Chem. Soc.* **2003**, *125*, 13670.

11. Semetey, V.; Rognan, D.; Hemmerlin, C.; Graff, R.; Briand, J. P.; Marraud, M.; Guichard, G. *Angew. Chem. Int. Ed.* **2002**, *41*, 1893.

12. Le Grel, P.; Salaun, A.; Potel, M.; Le Grel, B.; Lassagne, F. *J. Org. Chem.* **2006**, *71*, 5638.

13. Horne, W. S.; Gellman, S. H. *Acc. Chem. Res.* **2008**, *41*, 1399.

14. Guo, L.; Almeida, A. M.; Zhang, W.; Reidenbach, A. G.; Choi, S. H.; Guzei, I. A.; Gellman, S. H. *J. Am. Chem. Soc.* **2010**, *132*, 7868.

15. Miller, S. L. *J. Am. Chem. Soc.* **1955**, *77*, 2351.

16. Ehrenfreund, P., Glavin, D. P.; Botta, O.; Cooper, G.; Bada, J. L. *Proc. Natl. Acad. Sci. USA* **2001** *98*, 2138.

17. Dado, G. P.; Gellman, S. H. *J. Am. Chem. Soc.* **1994**, *116*, 1054.

18. Seebach, D.; Overhand, M.; Kuhnle, F. N. M.; Martinoni, B.; Oberer, L.; Hommel, U.; Widmer, H. *Helv. Chim. Acta* **1996**, *79*, 913.

63

19. Banerjee, A.; Balaram, P. *Curr. Science* **1997**, *73*, 1067.

20. Beke, T.; Somlai, C.; Perczel, A. *J. Comput. Chem.* **2006**, *27*, 20.

21. Appella, D. H.; Christianson, L. A.; Karle, I. L.; Powell, D. R.; Gellman, S. H. *J. Am. Chem. Soc.* **1996**, *118*, 13071.

22. Appella, D. H.; Barchi, J. J.; Durell, S. R.; Gellman, S. H. *J. Am. Chem. Soc.* **1999**, *121*, 2309.

23. Seebach, D.; Ciceri, P. E.; Overhand, M.; Jaun, B.; Rigo, D. *Helv. Chim. Acta* **1996**, *79*, 2043.

24. Seebach, D.; Abele, S.; Gademann, K.; Guichard, G.; Hintermann, T.; Jaun, B.; Matthews, J. L.; Schreiber, J. V.; Oberer, L.; Hommel, U.; Widmer, H. *Helv. Chim. Acta* **1998**, *81*, 932.

25. Seebach, D.; Schreiber, J. V.; Abele, S.; Daura, X.; Gunsteren, W. F. *Helv. Chim. Acta* **2000**, *83*, 34.

26. Hetényi, A.; Mándity, I. M.; Martinek, T. A.; Tóth, G. K.; Fülöp, F. *J. Am. Chem. Soc.* **2005**, *127*, 547.

27. Guichard, G.; Abele, S.; Seebach, D. *Helv. Chim. Acta* **1998**, *81*, 187.

28. Gung, B. W.; Zou, D.; Stalcup, A. M.; Cottrell, C. E. *J. Org. Chem.* **1999**, *64*, 2176.

29. Raguse, T. L.; Porter, E. A.; Weisblum, B.; Gellman, S. H. *J. Am. Chem. Soc.* **2002**, *124*, 12774.

30. Ebert, M. O.; Gardiner, J.; Ballet, S.; Abell, A. D.; Seebach, D. *Helv. Chim. Acta* **2009**, *92*, 2643.

31. Rueping, M.; Jaun, B.; Seebach, D. *Chem. Commun.* **2000**, 2267.

32. Arvidsson, P. I.; Rueping, M.; Seebach, D. *Chem. Commun.* **2001**, 649.

33. Cheng, R. P.; DeGrado, W. F. *J. Am. Chem. Soc.* **2001**, *123*, 5162.

34. Guarracino, D. A.; Chiang, H. J. R.; Banks, T. N.; Lear, D. J.; Hodsdon, M. E.; Schepartz A. *Org Lett.* **2006**, *8*, 807.

35. Christianson, L. A.; Lucero, M. J.; Appella, D. H.; Klein, D. A.; Gellman, S. H. *J. Comput. Chem.* **2000**, *21*, 763.

36. Appella, D. H.; Christianson, L. A.; Klein, D. A.; Powell, D. R.; Huang, X.; Barchi, J. J.; Gellman, S. H. *Nature* **1997**, *387*, 381.

37. Barchi, J. J.; Huang, X. L.; Appella, D. H.; Christianson, L. A.; Durell, A. R.; Gellman, S. H. *J. Am. Chem. Soc.* **2000**, *122*, 2711.

38. Appella, D. H.; Christianson, L. A.; Klein, D. A.; Richards, M. R.; Powell, D. R.; Gellman, S. H. *J. Am. Chem. Soc.* **1999**, *121*, 7574.

39. Wang, X.; Espinosa, J. F.; Gellman, S. H. *J. Am. Chem. Soc.* **2000**, *122*, 4821.

40. Lee, H.-S.; Syud, F. A.; Wang, X.; Gellman, S. H. *J. Am. Chem. Soc.* **2001**, *123*, 7721.

41. LePlae, P. R.; Fisk, J. D.; Porter, E. A.; Weisblum, B.; Gellman, S. H. *J. Am. Chem. Soc.* **2002**, *124*, 6820.

42. Park, J. S.; Lee, H. S.; Lai, J. R.; Kim, B. M.; Gellman, S. H. *J. Am. Chem. Soc.* **2003**, *125*, 8539.

43. Winkler, J. D.; Piatnitski, E. L.; Mehlmann, J.; Kasparec, J.; Axelsen, P. H. *Angew. Chem., Int. Ed.* **2001**, *40*, 743.

44. Hetényi, A.; Szakonyi, Z.; Mándity, I. M.; Szolnoki, É.; Tóth, G. K.;Martinek, T. A.; Fülöp, F. *Chem. Commun* **2009**, 177.

45. Hetényi, A.; Tóth, G. K.; Somlai, C.; Vass, E.; Martinek, T. A.; Fülöp, F. *Chem. Eur. J.* **2009**, *15*, 10736.

46. Gruner, S. A. W.; Truffault, V.; Voll, G.; Locardi, E.; Stöckle, M.; Kessler, H. *Chem. Eur. J.* **2002**, *8*, 4365.

47. Sharma, G. V. M.; Reddy, K. R.; Krishna, P. R.; Sankar, A. R.; Jayaprakash, P.; Jagannadh, B.; Kunwar, A. C. *Angew. Chem., Int. Ed.* **2004**, *43*, 3961.

48. Krauthauser, S.; Christianson, L. A.; Powell, D. R.; Gellman, S. H. *J. Am. Chem. Soc.* **1997**, *119*, 11719.

49. Chung, Y. J.; Christianson, L. A.; Stanger, H. E.; Powell, D. R.; Gellman, S. H. *J. Am. Chem. Soc.* **1998**, *120*, 10555.

50. Chung, Y. J.; Huck, B. R.; Christianson, L. A.; Stanger, H. E.; Krauthauser, S.; Powell, D. R.; Gellman, S. H. *J. Am. Chem. Soc.* **2000**, *122*, 3995.

51. Seebach, D.; Abele, S.; Gademann, K.; Jaun, B. *Angew. Chem. Int. Ed.* **1999**, *38*, 1595.

52. Daura, X.; Gademann, K.; Schäfer, H.; Jaun, B.; Seebach, D.; Gunsteren, W. F. *J. Am. Chem. Soc.* **2001**, *123*, 2393.

53. Martinek, T. A.; Tóth, G. K.; Vass, E.; Hollósi, M.; Fülöp, F. *Angew. Chem. Int. Ed.* **2002**, *41*, 1718.

54. De Pol, S.; Zorn, C.; Klein, C. D.; Zerbe, O.; Reiser, O. *Angew. Chem., Int. Ed.* **2004**, *43*, 511.

55. Hayen, A.; Schmitt, M. A.; Ngassa, F. N.; Thomasson, K. A.; Gellman, S. H. *Angew. Chem., Int. Ed.* **2004**, *43*, 505.

56. Sharma, G. V. M.; Nagendar, P.; Jayaprakash, P.; Krishna, P. R.; Ramakrishna, K. V. S.; Kunwar, A. C. *Angew. Chem., Int. Ed.* **2005**, *44*, 5878.

57. Srinivasulu, G.; Kumar, S. K.; Sharma, G. V. M.; Kunwar, A. C. *J. Org. Chem.* **2006**, *71*, 8395.

58. Jagadeesh, B.; Prabhakar, A.; Sarma, G. D.; Chandrasekhar, S.; Chandrashekar, G.; Reddy, M. S.; Jagannadh, B. *Chem.Commun.* **2007**, 371.

59. Schmitt, M. A.; Choi, S. H.; Guzei, I. A.; Gellman, S. H. *J. Am. Chem. Soc.* **2006**, *128*, 4538.

60. Wu, Y. D.; Wang, D. P. *J. Am. Chem. Soc.* **1998**, *120*, 13485.

61. Wu , Y. D.; Wang, D. P. *J. Am. Chem. Soc.* **1999**, *121*, 9352.

62. Günther, R.; Hofmann, H. J. *Helv. Chim. Acta* **2002**, *85*, 2149.

63. Ivánovics, G.; Bruckner, V. *Nuturwissenschaften* **1937**, *25*, 250.

64. Welch, B. D.; VanDemark, A. P.; Heroux, A.; Hill, C. P.; Kay, M. S. *Proc. Natl. Acad. Sci. USA* **2007** *104*, 16828.

65. Burkhart, B. M.; Gassma, R. M.; Langs, D A.; Pangborn, W. A.; Duax, W. L.; Pletnev, V. *Biopolymers* **1999**, *51*, 129.

66. Di Blasio, B.; Saviano, M.; Fattorusso, R.; Lombardi, A.; Pedone, C.; Valle, V.; Lorenzi, G. P. *Biopolymers* **1994**, *34*, 1463.

67. Raguse, T. L.; Lai, J. R.; LePlae, P. R.; Gellman, S. H. *Org. Lett.* **2001**, *3*, 3963.

68. Chakraborty, P.; Diederichsen, U. *Chem. Eur. J.* **2005**, *11*, 3207.

69. Lelais, G.; Seebach, D.; Jaun, B.; Mathad, R. I.; Flögel, O.; Rossi, F.; Campo, M.; Wortmann; A. *Helv. Chim. Acta* **2006**, *89*, 361.

70. Cheng, R. P.; DeGrado W. F. *J . Am. Chem. Soc.* **2002**, *124*, 11564.

71. Petersson, E. J.; Schepartz, A. *J . Am. Chem. Soc.* **2008**, *130*, 821.

72. Goodman, J. L.; Petersson, E. J.; Daniels, D. S.; Qiu, J. X.; Schepartz, A. *J . Am. Chem. Soc.* **2007**, *129*, 14746.

73. Horne, W. S.; Price, J. L.; Keck, J. L.; Gellman, S. H. *J. Am. Chem. Soc.* **2007**, *129*, 4178.

74. Choi, S. H; Guzei, I. A.; Spencer, L. C.; Gellman, S. H. *J. Am. Chem. Soc.* **2008**, *130*, 6544.

75. Pomerantz, W. C.; Yuwono, V. M.; Pizzey, C. L.; Hartgerink, J. D.; Abbott, N. L.; Gellman S. H. *Angew. Chem. Int. Ed.* **2008**, *47*, 1241.

76. Pomerantz, W. C.; Abbott, N. L.; Gellman S. H. *J. Am. Chem. Soc.* **2006**, *128*, 8730.

77. Martinek, T. A.; Hetényi, A.; Fülöp, L.; Mándity, I. M.; Tóth, G. K.; Dékány, I.; Fülöp, F. *Angew. Chem. Int. Ed.* **2006**, *45*, 2396.

78. Hintermann, T.; Seebach, D. *Chimia* **1997**, *50*, 244.

79. Seebach, D.; Abele, S.; Schreiber, J. V.; Martinoni, B.; Nussbaum, A. K.; Schild, H.; Schulz, H.; Hennecke, H.; Woessner, R.; Bitsch, F. *Chimia* **1998**, *52*, 734.

80. Oren, Z.; Shai, Y. *Biopolymers* **1998**, *47*, 451.

81. Tossi, A.; Sandri, L.; Giangaspero, A. *Biopolymers* **2000**, *55*, 4.

82. DeGrado, W. F.; Musso, G. F.; Lieber, M.; Kaiser, E. T.; Kezdy, F. J. *Biophys. J.* **1982**, *37*, 329.

83. Liu, D.; DeGrado, W. F. *J. Am. Chem. Soc.* **2001**, *123*, 7553.

84. Porter, E. A.; Wang, X.; Lee, H.-S.; Weisblum, B.; Gellman, S. H. *Nature* **2000**, *404*, 565.

85. Epand, R. F.; Raguse, T. L.; Gellman, S. H.; Epand, R. M. *Biochemistry* **2004**, **43**, 9527.

86. Brogden, K. A. *Nature Rev. Microbiol.* **2005**, *3*, 238.

87. Mowery, B. P.; Lee, S. E.; Kissounko, D. A.; Epand, R. F.; Epand, R. M.; Weisblum, B.; Stahl, S. S.; Gellman, S. H. *J. Am. Chem. Soc.* **2007**, *129*, 15474.

88. Epand, R. F.; Mowery, B. P.; Lee, S. E.; Stahl, S. S.; Lehrer, R. I.; Gellman, S. H.; Epand, R. M. *J. Mol. Biol.* **2008**, *379*, 38.

89. Hajduk, P. J.; Huth, J. R.; Tse, C. *Drug Disc. Today* **2005**, *10*, 1675.

90. Werder, M.; Hauser, H.; Abele, S.; Seebach, D. *Helv. Chim. Acta* **1999**, *82*, 1774.

91. Gelman, M. A.; Richter, S.; Cao, H.; Umezawa, N.; Gellman, S. H.; Rana, T. M. *Org. Lett.* **2003**, *5*, 3563.

92. Rueping, M.; Mahajan, Y.; Sauer, M.; Seebach, D. *ChemBioChem* **2002**, *3*, 257.

93. Potocky, T. B.; Menon, A. K.; Gellman, S. H. *J. Biol. Chem.* **2003**, *278*, 50188.

94. Eldred, S. E.; Pancost, M. R.; Otte, K. M.; Rozema, D.; Stahl, S. S.; Gellman, S. H. *Bioconj. Chem.* **2005**, *16*, 694.

95. Kritzer, J.; Lear, J. D.; Hodsdon, M. E.; Schepartz, A. *J. Am. Chem. Soc.* **2004**, *126*, 9468.

96. Kritzer, J. A.; Hodsdon, M. E.; Schepartz, A. *J. Am. Chem. Soc.* **2005**, *127*, 4118.

97. Chene, P. *Nat.Rev. Cancer* **2003**, *3*, 102.

98. Stephens, O.; Kim, S.; Welch, B. D.; Hodsdon, M. E.; Kay, M. S.; Schepartz, A. *J. Am. Chem. Soc.* **2005**, *127*, 13126.

99. Kritzer, J. S.; Stephens, O. M.; Guarracino, D. A.; Reznik, S. K.; Schepartz, A. *Bioorg. Med. Chem.* **2005**, *13*, 11.

100. Horne, W. S.; Johnson, L. M.; Ketas, T. J.; Klasse, P. J.; Lu, M.; Moore, J. P.; Gellman, S. H. *Proc. Natl. Acad. Sci. USA* **2009**, *106*, 14751.

101. Imamuram, Y.; Watanabe, N.; Umezawa, N.; Iwatsubo, T.; Kato, N.; Tomita, T.; Higuchi, T. *J. Am. Chem. Soc.* **2009**, *131*, 7353.

102. Nunn, C.; Langenegger, M. R. D.; Schuepbach, E.; Kimmerlin, T.; Micuch, P.; Hurth, K.; Seebach, D.; Hoyer, D. *Naunyn-Schmiedeberg's Arch. Pharmacol.* **2003**, *367*, 95.

103. Wiegand, H.; Wirz, B.; Schweitzer, A.; Gross, G.; Rodriguez-Perez, M. I.; Andres, H.; Kimmerlin, T.; Rueping, M.; Seebach, D. *Chem. Biodiversity* **2004**, *1*, 1812.

104. Trouche, N.; Wieckowski, S.; Sun, W.; Chaloin, O.; Hoebeke, J.; Fournel, S.; Guichard, G. *J. Am. Chem. Soc.* **2007**, *129*, 13480.

105. Fournel, S.; Wieckowski, S.; Sun, W.; Trouche, N.; Dumortier, H.; Bianco, A.; Chaloin, O.; Habib, M.; Peter, J.-C.; Schneider, P.; Vray, B.; Toes, R. E.; Offringa, R.; Melief, C. J. M.; Hoebeke, J.; Guichard, G. *Nat. Chem. Biol.* **2005**, *1*, 377.

106. Weiss, H. M.; Wirz, B.; Schweitzer, A.; Amstutz, R.; Rodriguez-Perez, M. I.; Andres, H.; Metz, Y.; Gardiner, J.; Seebach, D. *Chem. Biodiversity* **2007**, *4*, 1413.

107. Wiegand, H.; Wirz, B.; Schweitzer, A.; Camenisch, G. P.; Rodriguez Perez, M. I.; Gross, G.; Woessner, R.; Voges, R.; Arvidsson, P. I.; Frackenpohl, J.; Seebach, D. *Biopharm. Drug Dispos.* **2002**, *23*, 251.

108. Merrifield, R. B. *J. Am. Chem. Soc.* **1963**, *85*, 2149.

109. Carpino, L. A. *Acc. Chem. Res.* **1987**, *20*, 401.

110. Kaiser, E.; Colescott, R. L.; Bossinger C. D.; Cook, P. *J. Anal. Biochem.* **1970**, *34*, 595.

111. Murray, J. K.; Gellman, S. H. *Org. Lett.* **2005**, *7*, 1517.

112. Pietta, P. G.; Cavallo, P. F.; Takahashi, K.; Marshall, G. R. *J. Org. Chem.* **1974**, *39*, 44.

113. Rink, H. *Tetrahedron Lett.* **1987**, *28*, 3787.

114. Fuller, W. D.; Goodman, M.; Naider, F. R.; Zhu, Y. F. *Biopolymers* **1996**, *40*, 183.

115. Seebach, D.; Kimmerlin, T.; Šebesta, R.; Campo, M. A.; Beck, A. K. *Tetrahedron* **2004**, *60*, 7455.

116. Chan, W. C.; White, P. D. (Eds.) Fmoc Solid Phase Peptide Synthesis: A Practical Approach *Oxford University Press: Oxford*, **2000**.

117. Kimmerlin, T.; Seebach, D.; Hilvert, D. *Helv. Chim. Acta* **2002**, *85*, 1812.

118. Ingenito, R.; Bianchi, E.; Fattori, D.; Pessi, A. *J. Am. Chem. Soc* **1999**, *121*, 11369.

119. Backes, B. J.; Ellman, J. A. *J. Org. Chem.* **1999**, *64*, 2322.

120. http://www.mn-net.com/tabid/6121/default.aspx

121. http://www.knauer.net/

122. http://www.home.agilent.com/agilent/home.jspx?cc=US&lc=eng

123. http://www.phenomenex.com/

124. http://www.thermoscientific.com/

125. Bayer, E. *Angew. Chem. Int. Ed.* **1991**, *30*, 113.

126. Carpino, L. A. *J. Am. Chem. Soc.* **1993**, *115*, 4379.

127. http://www.jascoinc.com/Home.aspx

128. http://www.thalesnano.com/

129. http://www.bruker-biospin.com/nmr.html

130. Braunschweiler, L.; Ernst, R. R. *J Magn. Reson.* **1983**, *53*, 521.

131. Bax, A.; Davis, D. G. *J Magn. Reson.* **1985**, *65*, 355.

132. Bax, A.; Davis, D. G. *J Magn. Reson.* **1985**, *63*, 207.

133. Jeener, J.; Meier, B. H.; Bachmann, P.; Ernst, R. R. *J. Phys. Chem.* **1979**, *69*, 4546.

134. Wagner, G.; Wüthrich, K. *J. Mol Biomol.* **1982**, *155*, 347.

135. Croasmun, W. R.; Carlson, R. M. K (Eds.). Two-dimensional NMR spectroscopy application for chemists and biochemists second edition. VCH, New York, **1994**.

136. http://www.bruker-biospin.com/nmr_software.html

137. http://www.jascoinc.com/Products/Spectroscopy/J-815-Circular-Dichroism-Spectrometer.aspx

138. http://www.jasco.co.uk/Spectra_Manager_II.asp

139. http://www.chemcomp.com

140. Halgren, T. A. *J. Comput. Chem.* **1996**, *17*, 490.

141. Frisch, M. J.; Trucks, G. W.; Schlegel, H. B.; Scuseria, G. E.; Robb, M.A.; Cheeseman, J. R.; Montgomery, J. A.; Jr.; Vreven, T.; Kudin, K. N.; Burant, J.

C.; Millam, J. M.; Iyengar, S. S.; Tomasi, J.; Barone, V.; Mennucci, B.; Cossi, M.; Scalmani, G.; Rega, N.; Petersson, G. A.; Nakatsuji, H.; Hada, M.; Ehara, M.; Toyota, K.; Fukuda, R.; Hasegawa, J.; Ishida, M.; Nakajima, T.; Honda, Y.; Kitao, O.; Nakai, H.; Klene, M.; Li, X.; Knox, J. E.; Hratchian, H. P.; Cross, J. B.; Adamo, C.; Jaramillo, J.; Gomperts, R.; Stratmann, R. E.; Yazyev, O.; Austin, A. J.; Cammi, R.; Pomelli, C.; Ochterski, J. W.; Ayala, P. Y.; Morokuma, K.; Voth, G. A.; Salvador, P.; Dannenberg, J. J.; Zakrzewski, V. G.; Dapprich, S.; Daniels, A. D.; Strain, M. C.; Farkas, O.; Malick, D. K.; Rabuck, A. D.; Raghavachari, K.; Foresman, J. B.; Ortiz, J. V.; Cui, Q.; Baboul, A. G.; Clifford, S.; Cioslowski, J.; Stefanov, B. B.; Liu, G.; Liashenko, A.; Piskorz, P.; Komaromi, I.; Martin, R. L.; Fox, D. J.; Keith, T.; Al-Laham, M. A.; Peng, C. Y.; Nanayakkara, A.; Challacombe, M.; Gill, P. M. W.; Johnson, B.; Chen, W.; Wong, M. W.; Gonzalez, C.; Pople, J. A. Gaussian 03, Revision A.1; Gaussian, Inc.: Pittsburgh, PA, 2003; http://www.gaussian.com.

142. Tomasi, J.; Mennucci, B.; Cammi, R. *Chem. Rev.* **2005,** *105,* 2999.

143. Engelhaaf, S. U.; Lobaskin, V.; Bauer, H. H.; Nerkle, H. P.; Schurtenberger, P. *Eur. Phys. J. E.* **2004,** *13,* 153.

144. Valluzi, R.; Kaplan, D. L. *Biopolymers* **2000,** *53,* 350; Forró, E.; Fülöp, F. *Chem. Eur. J.* **2007,** *13,* 6397.

145. Baldauf, C.; Günther, R.; Hofmann, H. J. *Biopolymers* **2005,** *80,* 675.

146. Molski, M. A.; Goodman, J. L.; Craig, C. J.; Meng, H.; Kumar, K.; Schepartz, A. *J. Am. Chem. Soc.* **2010,** *132,* 3658.

147. Chandrasekhar, S.; Sudhakar, A.; Kiran, M. U.; Babu, B. N.; Jagadeesh, B. *Tetrahedron Lett.* **2008,** *49,* 7368.

148. Mohle, K.; Gunther, R.; Thormann, M.; Sewald, N.; Hofmann, H. J. *Biopolymers* **1999,** *50,* 167.

149. Beke, T.; Csizmadia, I G.; Perczel, A. *J. Comput. Chem.* **2004,** *25,* 285.

150. Topol, I. A.; Burt, S. K.; Deretey, E.; Tang, T. H.; Perczel, A.; Rashin, A.; Csizmadia I. G. *J. Am. Chem. Soc.* **2001,** *123,* 6054.

151. Townsley, L. E., Tucker, W. A., Sham, S.; Hinton, J. F. *Biochemistry* **2001**, *40*, 11676.

152. Claridge, T. D. W.; Goodman, J. M.; Moreno, A.; Angus, D.; Barker, S. F.; Taillefumier, C.; Watterson, M. P.; Fleet, G. W. J. *Tetrahedron Lett.* **2001**, *42*, 4251.

153. Abele, S.; Seebach, D. *Helv. Chim. Acta* **1999**, *82*, 1559.

154. Ramachandran, G. N.; Ramakrishnan, C.; Sasisekharan, V. *J. Mol. Biol.* **1963**, *7*, 95.

155. Wu, Y. D.; Han, W.; Wang, D. P.; Gao, Y.; Zhao, Y. L. *Acc. Chem. Res.* **2008**, *41*, 1418.

156. Toniolo, C.; Crisma, M.; Formaggio, F.; Peggion, C. *Biopolymers* **2001**, *60*, 396.

157. Toniolo, C.; Bonora, G. M.; Stavropoulous, G.; Cordopatis, P.; Theodoropoulos, D. *Biopolymers* **1986**, *25*, 281.

I. Martinek, T. A.; Mándity, I. M.; Fülöp, L.; Tóth, G. K.; Vass, E.; Hollósi, M.; Forró, E.; Fülöp, F. *J. Am. Chem. Soc.* **2006**, *128*, 13539.

II. Mándity, I. M.; Wéber, E.; Martinek, T. A.; Olajos, G.; Tóth, G. K.; Vass, E.; Fülöp, F. *Angew. Chem. Int. Ed.* **2009**, *48*, 2171.

III. Mándity, I. M.; Fülöp, L.; Vass, E.; Tóth, G. K.; Martinek, T. A.; Fülöp, F. *Org. Lett.*, **2010**, *12*, 5584.

Printed by Books on Demand GmbH, Norderstedt / Germany